Radical Reactions of Fullerenes and their Derivatives

Developments in Fullerene Science

Volume 2

Series Editor:

Tibor Braun, *Institute of Inorganic and Analytical Chemistry, L. Eötvös University, Budapest, Hungary*

The titles published in this series are listed at the end of this volume.

Radical Reactions of Fullerenes and their Derivatives

by

Boris Tumanskii
Nesmeyanov Institute of Organoelement Compounds, Moscow, Russia

and

Oleg Kalina
Nesmeyanov Institute of Organoelement Compounds, Moscow, Russia

SPRINGER-SCIENCE+BUSINESS MEDIA, B.V.

A C.I.P. Catalogue record for this book is available from the Library of Congress.

ISBN 978-90-481-5892-8 ISBN 978-0-306-47634-1 (eBook)
DOI 10.1007/978-0-306-47634-1

Printed on acid-free paper

Contents

Contributors

The authors of this monograph, Boris L. Tumanskii and Oleg G. Kalina, work in the Nesmeyanov Institute of Organoelement Compounds (INEOS) in Moscow, Russian Federation.

Boris L. Tumanskii received his Doctor's degree in 1984 from INEOS. He was promoted to Professor in 1996. His research interests are in the area of radical chemistry. Last 10 years were devoted to studies of radical reactions of fullerenes and their derivatives. He is the author of over 200 scientific publications connected with the EPR study of the structure and reactivity of free radicals.

Oleg G. Kalina received his Master's degree from Higher Chemical College of Russian Academy of Sciences in 1999, and Doctor's degree from INEOS in 2001. He has held the Soros Science Education Program Award in 2000 and 2001. His research interests are the synthesis of new fullerene derivatives, quantum-chemical calculations of paramagnetic compounds, and EPR spectroscopy.

Acknowledgements

The authors are grateful to the Russian Foundation for Basic Research (projects 01-03-33216 and 99-03-33067), the State Science and Engineering Program "Topical Lines of Research in the Physics of Condensed Media, Line: Fullerenes and Atomic Clusters," and to the subprogram "Fundamental Problems of Modern Chemistry" (project 9.4.06).

We are indebted to our co-workers: V. V. Bashilov, S. P. Solodovnikov, N. N. Bubnov, Yu. I. Lyakhovetsky, E. A. Shilova, V. I. Sokolov, R. G. Gasanov, V. I. Stankevich, A. L. Chistyakov, Yu. N. Novikov.

We thank I. Yu. Kudryavtsev for his help in checking English.

Oleg Kalina would like to express his deep appreciation to Tatyana Pankova for her love and support.

Preface

The discovery of fullerenes, species belonging to the electronodeficient polyalkenes with weakly conjugated double bonds, has opened novel opportunities for the radical chemistry. Pioneering study in this field was performed by P. J. Krusic, E. Wasserman, P. N. Keizer, J. R. Morton, and K. F. Preston (*Science*, 1991, **254**, 1184).

The fullerenyl radical adducts formed *via* addition of atoms or free radicals to fullerenes have no analogs in organic chemistry. In fact, radicals in which the unpaired electrons are delocalized over the surface of a sphere or ellipsoid have never been studied before. The unusual character of the fullerenyl radicals is also due to the fact that they occupy a sort of intermediate position between the planar π-radicals and tetrahedral σ-radicals. Thus, the elucidation of the characteristic features of fullerenyl radicals and their reactivity by EPR spectroscopy, and the comparison of the results with those of quantum-chemical studies are of fundamental importance.

Isolation of the products from homolytic reactions of fullerenes in bulk amounts opens the door to large-scale preparation of new organic and organoelement derivatives of C_{60} including biologically active ones. Radical reactions of fullerenes find wide application in the synthesis of fullerene-containing polymers with valuable photophysical characteristics.

Ferromagnetism of the complex of C_{60} with tetra(dimethylamino)ethylene found lends impetus to a search of novel methods for preparation of biradicals one unpaired electron of those is located on the fullerene cage while the other retained by the addend.

The capacity of [60]fullerene to add a great number of radicals that is also characteristic of its derivatives soluble in water (e. g. $C_{60}(OH)_n$) has been already employed in order to effectively inhibit radical processes occurring in living cells.

Given such various and promising properties of fullerene derivatives, creation of new methods for chemical functionalization of fullerenes including those of radical chemistry is a matter of current interest.

This book covers the results of studies on homolytic reactions of fullerenes, the structures and reactivities of fullerenyl radicals performed by EPR, quantum-chemical, and other methods. The attention is focused upon the following questions:

1. The reactivity of fullerenes towards atoms and radicals of different chemical structures;
2. Structural features of fullerenyl radicals;
3. The impact of the unpaired electron delocalization upon the reactivities of fullerenyl radicals;
4. Criteria defining the regiochemistry of radical particles addition to the higher fullerenes containing non-equivalent carbon atoms;
5. The reactivities of non-equivalent vertices of the fullerene carbon framework in organic derivatives of fullerene towards attacking radicals;
6. Structures of adducts of multiple addition of radical particles or atoms to fullerene;
7. Manifestations of the magnetic interaction between two unpaired electrons located on fullerene and the attached organic fragment correspondingly;
8. Outlook. The further development of the fullerene radical chemistry.

Examples of the application of radical reactions of fullerenes in the synthesis and typical methods for their investigation are given. In order to enlarge the circle of readers, both a brief introduction into the EPR method - a main tool for the exploration of free radicals, and general information about free radicals, methods of their generation and transformations are also included.

Chapter 1

Basic Data on Fullerenes and Free Radicals

1.1 Features of Fullerenes

Fullerenes are carbon clusters,[1] whose surface is formed by twelve pentagons and any number of hexagons. The majority of studies of fullerenes have involved C_{60} (Fig. 1), which is the most abundant compound in the process of fullerene generation.

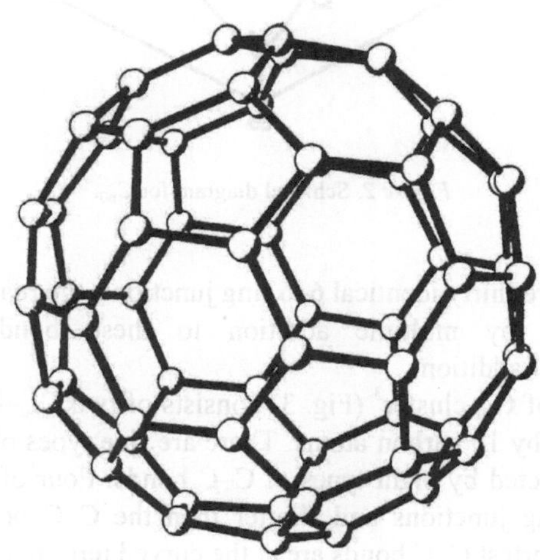

Figure 1. C_{60} fullerene.

Fullerene C_{60} consists of sixty equivalent carbons linked by two types of C–C bonds (Fig. 2). The bonds at the 6–6 ring junctions are shorter (bond length 1.38 Å) than the bonds at the 6–5 ring junctions (bond length 1.45 Å).[2] Therefore, C_{60} is not so much an aromatic molecule as giant closed-cage alkenes and is much more reactive than expected.

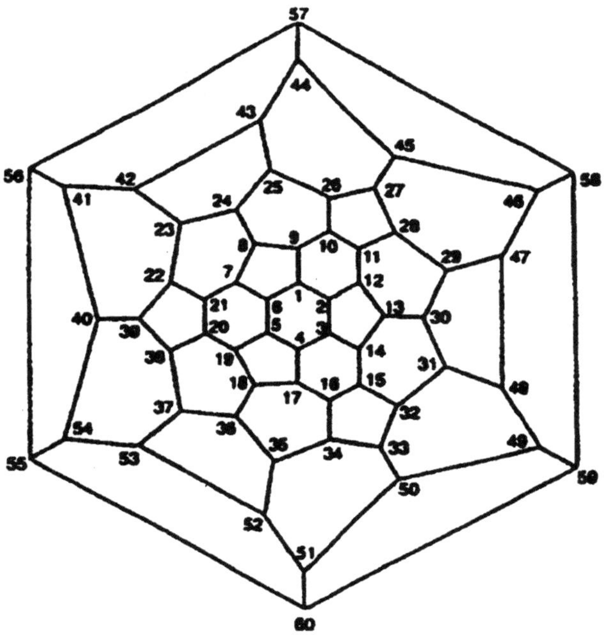

Figure 2. Schlegel diagram for C_{60}.

Since there are thirty identical 6–6 ring junctions, the reactions of C_{60} can be complicated by multiple addition to these bonds and by the regiochemistry of addition.

A molecule of C_{70} cluster[3] (Fig. 3) consists of two C_{60}–like hemispheres that are bridged by 10 carbon atoms. There are five types of carbons in C_{70}, which are connected by eight types of C–C bonds. Four of these bonds are between 6–6 ring junctions and shorter than the C–C bonds at 6–5 ring junction. The shortest C–C bonds are at the curved termini of the molecule.

Cluster C_{76} (D_{2d}) (Fig. 4) comprises 30 non-equivalent sets of 114 bonds, with each set containing either two or four identical bonds. The data show the general trend of the bonds with the highest π-density to occupy exocyclic

position in the pentagons. Addition is anticipated to take place in the vicinity of the polar caps.[4]

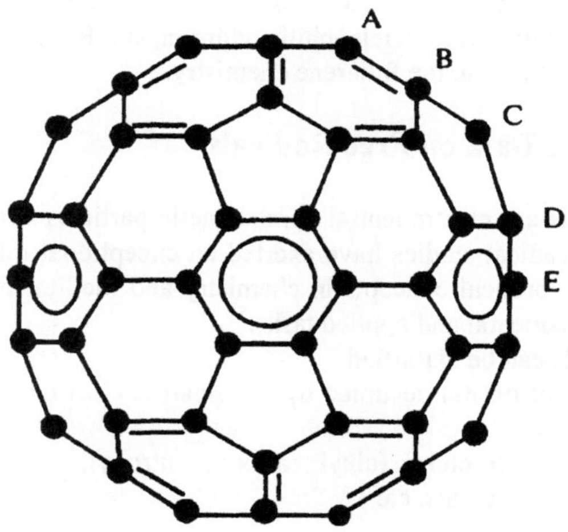

Figure 3. C_{70} fullerene. A-E are non-equivalent carbon atoms.

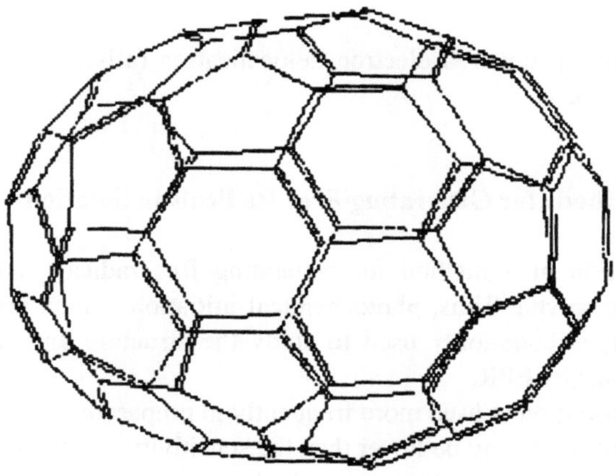

Figure 4. C_{76} fullerene.

Five isomeric structures are possible for C_{78} and 24 isomers for C_{84}.[5]

Because all fullerenes consist of only sp^2-hybridized atoms, they are electron-withdrawing molecules reacting as unsaturated polyalkenes in chemical transformations, such as cycloaddition, η^2-addition of organometallic derivatives, nucleophilic addition, etc. Radical reactions also play an important role in the fullerene chemistry.

1.2 Basic Data on Free Radicals

Free radicals are electroneutral paramagnetic particles with an unpaired electron. Free radical studies have exerted an exceptional influence on the evolution of theoretical concepts in chemistry and facilitated solution of a number of fundamental and applied tasks.[6]

Free radicals can be classified:

1. By the type of orbital occupied by an unpaired electron, for example, σ and π radicals;
2. By chemical structure (alkyl, aroxyl, nitroxyl, organophosphorus, organometallic radicals, etc.):

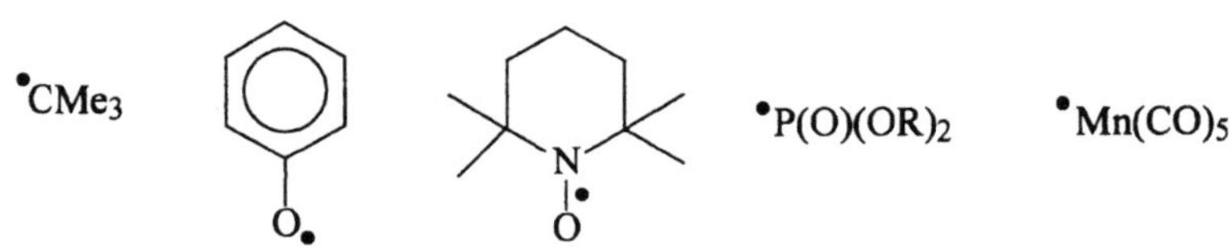

3. By the type of unpaired electron delocalization (allyl, cyclopentadienyl radicals, etc.).

1.2.1 Methods for Generating Free Radicals in Solution

The selection of a method for generating free radicals depends on a problem to be solved. Thus, photochemical initiation, which provides high initiation rate, is commonly used to study the structure and reactivity of unstable radicals by EPR.

Thermal initiation is used more frequently in preparative reactions where radical concentration may be lower than the sensitivity of EPR spectrometer.

We will focus our attention on methods successfully used to study the radical reactions of fullerenes. Systems producing radicals of one type upon photolysis or thermolysis were found to be the most convenient.

1.2.1.1 Cleavage of R-X and R-R Bonds as a Result of Photolysis or Thermolysis

Photolysis of organomercury compounds is one of important methods for preparing radicals through R–X bond cleavage.[7-14] Both carbon-centered and element-centered radicals can be obtained by this method:

$$R{-}Hg{-}R \xrightarrow{h\nu} 2\,{}^{\bullet}R + Hg$$

R = $C(CH_3)_3$, $CH(CH_3)_2$, CH_2Ph, Ph, C(O)Me, $C(CF_3)_3$, $CF{=}CF(CF_3)_3$, SAlk, $P(O)(OX)_2$ (where X = Me, Et, Pr^i); $B_{10}H_9C_2H_2$-*o,m,p*, $Pt(PPh_3)_2CHPh_2$, etc.

Alkyl radicals were also obtained by photolysis of alkyl iodides or bromides:[15,16]

$$Alk{-}X \xrightarrow{h\nu} {}^{\bullet}Alk + {}^{\bullet}X,\ X = I, Br;\ Alk = C(CH_3)_3,\ CF_2CF_3$$

Photolysis of polycarbonyl compounds was used to produce acyl radicals:[17]

$$(CF_3)_2CF(CO)(CO)(CO)CF(CF_3)_2 \xrightarrow{h\nu} 2\,{}^{\bullet}C(O)CF(CF_3)_2 + CO$$

Alkoxy radicals can be photolytically generated from their peroxides. Dialkoxy disulfides ROSSOR are more convenient source of alkoxy radicals for addition reactions under photolytic conditions. These compounds can be easily prepared.[18]

Irradiation of methyl disulfide, MeSSMe leads to the formation of S-centered radicals.[14]

Alkoxyamines, the adducts of stable nitroxyl radical TEMPO (2,2,6,6,-tetramethylpiperidinyl-1-oxy) to different benzyl-type radicals, were employed as promising radical sources for introducing a definite number of radicals (including polymeric ones) to fullerene.[19]

These adducts dissociate at the O–C bond at about +125 to +140 °C to give the stable nitroxyl radical and reactive radical, which undergoes addition to fullerene:[19]

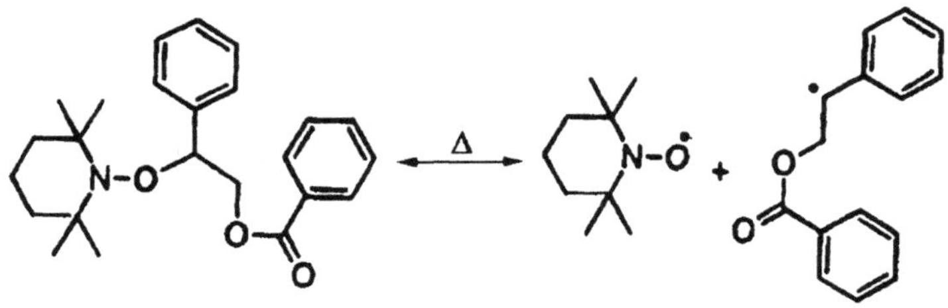

The examples of preparation of fullerene-containing polymers by this procedure will be described in Chapter 7.

The methods of preparing halogen atoms for the preparative synthesis of polyhalogenated products of addition to fullerene are also based on bond photolysis (e.g., Cl–I, Br–Br bonds).

Selective fluorination is accomplished with the use of heterogeneous systems where fluorine atoms result from the cleavage of metal–fluorine bonds.[3,20]

The photolysis of SF_5Cl was employed in the EPR studies of monoadducts of Cl and F atoms to fullerene.[21]

1.2.1.2 Monovalent Residue Abstraction by Another Radical

Hydrogen atom abstraction was commonly carried out with the use of radicals formed in the photolysis of *tert*-butyl peroxide:[22-24]

$$(CH_3)_3C{-}O{-}O{-}C(CH_3)_3 \xrightarrow{h\nu} 2\,{}^{\bullet}O{-}C(CH_3)_3$$

$$RH + {}^{\bullet}O{-}C(CH_3)_3 \longrightarrow {}^{\bullet}R + HO{-}C(CH_3)_3$$

$R = CH_2Ph, CH(CH_3)_2, P(O)(OMe)_2, BH_2NMe_3$, etc.

A large excess of RH with respect to the peroxide is used to prevent the addition of *tert*-butyl radicals or their fragmentation products, methyl radicals, to fullerene.

Certain carbon-centered radicals were also obtained by halogen abstraction from appropriate organic halides by metal-centered radicals formed upon photolysis of compounds containing metal–metal bond:[25]

$$(CO)_5Re\text{–}Re(CO)_5 \xrightarrow{h\nu} 2\ {}^{\bullet}Re(CO)_5$$

$$RHal + {}^{\bullet}Re(CO)_5 \longrightarrow {}^{\bullet}R + HalRe(CO)_5$$

Hexamethyldistannane was used as iodine abstraction initiator for the preparative synthesis of products of perfluoroalkyl radical addition to fullerene:[26]

$$(CH_3)_3Sn\text{–}Sn(CH_3)_3 \xrightarrow{h\nu} 2\ {}^{\bullet}Sn(CH_3)_3$$

$$R_FI + {}^{\bullet}Sn(CH_3)_3 \longrightarrow {}^{\bullet}R_F + ISn(CH_3)_3$$

It was also shown that radical p-$FC_6H_4(CF_3)_2C^{\bullet}$ (Fig. 5) can abstract hydrogen atoms from various organic molecules. The radical is readily formed at room temperature upon dissociation of dimer $[C(CF_3)_2C_6H_4F]_2$:[27,28]

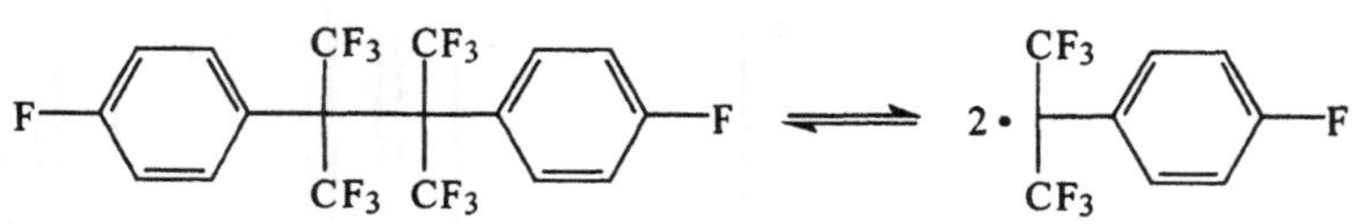

Hydrogen abstraction from toluene, mesitylene, hydrogen phosphites $HP(O)(OEt)Ph$, $HP(O)(OEt)_2$, $HP(O)(OPr^i)_2$, and aminoboryl compounds H_3BNMe_3 was shown to be possible. However, hydrogen cannot be detached from the methyl group in the *tert*-butylbenzene molecule. This seems to be due to the fact that the energy of the C–H bond in the methyl group in *tert*-butylbenzene (95 kcal mol^{-1}) is greater than that in toluene (83 kcal mol^{-1}). Thus, this dimer could be used for the generation of radicals from compounds in which C–H bond energy does not exceed 85 kcal mol^{-1}.[27]

A study of the reaction of the radical with C_{60} showed that no addition to C_{60} occurred without a solvent (below 430 K) or in a saturated benzene solution (350 K). In both cases, the EPR spectra exhibited only the signal for the radical itself with a characteristic hyperfine coupling of the unpaired electron with the fluorine and hydrogen nuclei (Fig. 5). The absence of the reaction between C_{60} and the radical seems to be due to steric shielding of the radical center. We will show bellow that the application of this radical as initiator makes it possible to observe the sequential addition of benzyl-type radicals to C_{60} and its derivatives.

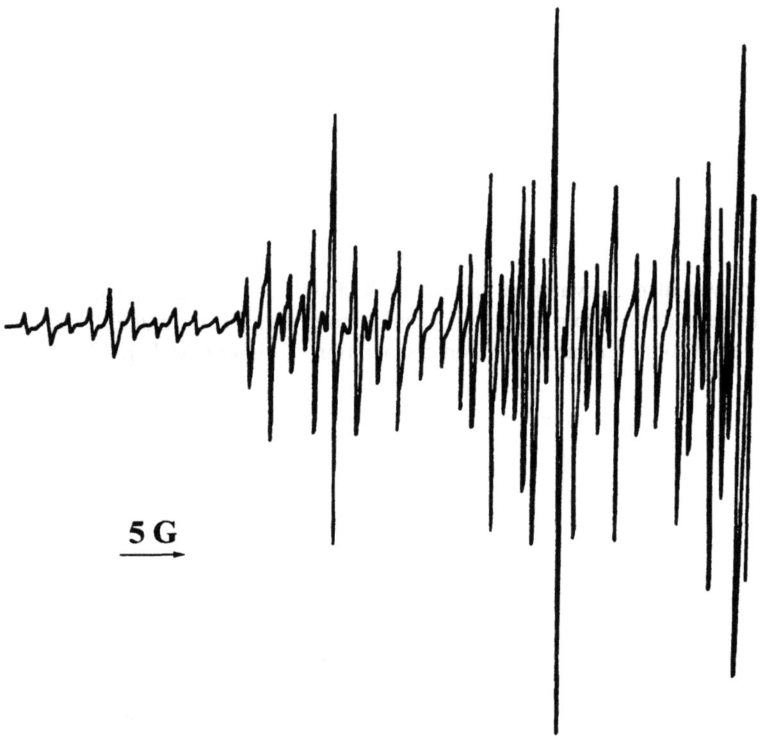

Figure 5. A half of EPR spectrum of $FC_6H_4(CF_3)_2C^{\bullet}$.

1.2.1.3 Generation of Free Radicals via Reactions Catalyzed by Electron Transfer

Electron transfer provides a basis for preparing the products of hydrogen atom addition to fullerene. The photochemical[29] or electrochemical[30,31] reduction of fullerene at the first stage brings about the formation of fullerene anion radical whose subsequent protonation leads to the product of hydrogen atom addition to fullerene:

$$C_{60} + e \longrightarrow C_{60}^{-\bullet}$$

$$C_{60}^{-\bullet} + H^{+} \longrightarrow {}^{\bullet}C_{60}H$$

The use of different catalytic systems allows radical generation at lower temperatures than in simple thermal homolysis. This is caused by lower activation energy of radical formation in processes where rate-determining step is one-electron transfer rather than the thermal dissociation of a covalent bond. This method was applied to preparative experiments. Thus, fullerene-containing polymers were obtained using monovalent copper complex CuBr/bipy as the catalyst of Br atom abstraction.[32]

1.2.2 Reactions Occurring via Free Radicals

Reactions involving unpaired electron are characteristic of all free radicals. These reactions are recombination, disproportionation, substitution, addition, and fragmentation.

Since the ability of free radicals to undergo addition to alkenes is the most interesting for the radical chemistry of fullerenes, let us consider this reaction in detail:

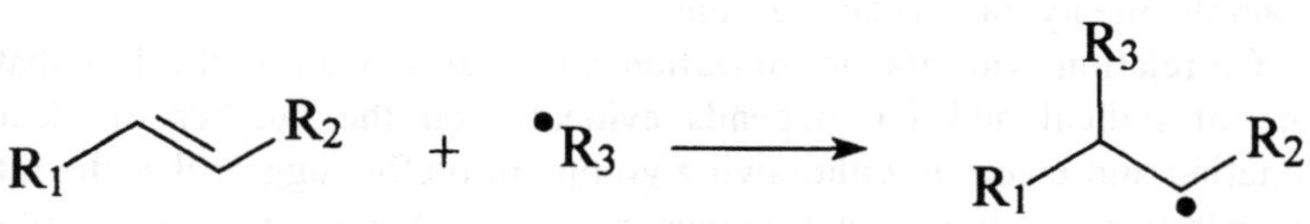

The addition of the simplest alkyl radicals—such as methyl, trifluoromethyl, trichloromethyl, and ethyl—to alkenes was studied in the most detail. The results of study on the direction of addition of trifluoromethyl radicals to unsymmetrical alkenes were reported in the work.[33]

Four factors were used to explain the direction of the radical addition reactions:[6]

1. The strength of resultant bond. Radical attaches itself to the position where the strongest bond forms;
2. Steric hindrances. Radical avoids spatially shielded sites either because the approach of the radical from this direction is sterically hindered or because the nascent bond in this case is to be weakened by steric interactions in the adduct formed;

3. Electrostatic effects. Radicals containing, e.g., halogens have to be polarized and exert electrophilic properties. If alkene comprises electron-withdrawing substituents, electrostatic interactions will arise between dipoles of the radical and the alkene, and the radical will bind to the carbon at the double bond that possesses lower positive charge;
4. The stability of radical adduct. Radical reacts with olefin in such a manner that the most stable product forms. The stability of radical adduct is commonly assessed from the delocalization of unpaired electron within the adduct.

As a rule, different combinations of all noted factors have to be taken into consideration to explain the regiochemistry of radical additions to alkenes.

Transition state is very important for understanding the regiochemistry of radical additions to alkenes.[6] If the transition state more resembles the initial species rather than the reaction products, then the properties of the starting compounds will have the greatest influence on the direction and rate of the reaction. Electrostatic and steric interactions between the radical and alkene should prevail in this case, whereas the stability of the final product is of minor significance.

If the transition state is closer in structure to the reaction product, then the stability of the resultant adduct will be the dominant factor. The selection of model seems to be dependent on alkene structure.

Thus, electrostatic and steric factors prevail for alkenes containing no functional groups capable of efficient delocalization of unpaired electron, for example, phenyl or carbonyl groups.[6]

Correlation with alkene ionization potential as well as the fact that the rate of radical addition depends evidently on the number of electron-donating and electron-withdrawing groups in olefin suggest that the rate of the addition reaction is determined by double bond properties, while the positioning of the radical at this or that carbon atom is accomplished in a latter reaction step. The activation energies for the addition of $^{\bullet}CCl_3$ and $^{\bullet}C_3F_7$ radicals to unsymmetrical alkenes were shown to be different for each terminus of the double bond. Consequently, in the rate-determining step, the radical undergoes addition to a specific atom of the alkene rather than to the double bond as a whole followed by migration of the radical to one of double bond termini.[6]

The association of radical with one specific carbon atom of double bond leads to so-called σ-transition state (Fig. 6). This state also resembles the initial species, and the arising carbon–carbon bond is longer. The attack at each of the atoms connected by double bond gives a particular σ-transition state. The activation energy of the addition is the difference between the energy of the highest point of potential curve and the energy of the initial species. Since the activation energies experimentally determined for $^{\bullet}CCl_3$

and $^{\bullet}C_3F_7$ differ when the radicals combine with diverse atoms linked by double bond in unsymmetrical olefins, the σ-transition state should have higher energy level than the transition state of the π-complex for the same radical.[6]

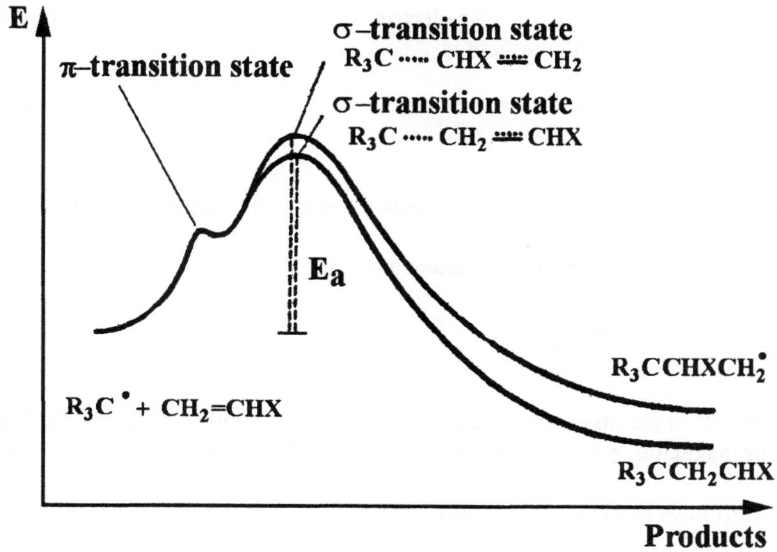

Figure 6. Potential energy diagram for addition reaction of alkyl radicals (X = F). Adapted from ref 6.

1.3 EPR Fundamentals

The study of paramagnetic species formed in radical reactions is possible due to EPR spectroscopy.[34]

Electron paramagnetic resonance (EPR), one of the most outstanding achievements of the 20th century, was discovered by Russian physicist Zavoisky.

EPR allows direct monitoring of unpaired electron behavior, that is, paramagnetism of free radicals, ions of variable valence, and of triplet molecules. That is why EPR enables the most delicate details of the structures and properties to be studied.

1.3.1 EPR Theory

A paramagnetic species containing an unpaired electron, when placed in a uniform magnetic field, **H**, directed along the **Z** axis, will have its spin and electron magnetic moment oriented either in parallel or in antiparallel with the magnetic field direction. The electron energy due to the interaction of the magnetic moment with the field is

$$E = -\mu_z H \quad (1)$$

where μ_z is the projection of the magnetic moment onto the field direction.

The magnetic moment component along the **Z** axis, μ_z, is

$$\mu_z = \mu M_s/S = -g\beta M_s \quad (2)$$

where $\mathbf{M_s}$ is the magnetic quantum number equal to either +1/2 or -1/2, β is Bohr magneton, **g** is a proportionality coefficient usually termed the g-factor.

It follows from (1) and (2) that

$$E = g\beta M_s H \quad (3)$$

Since the spin orientation in the magnetic field allows $\mathbf{M_s}$ to have just two values (±1/2) the electron energy in the field is

$$E = +1/2\ g\beta H \text{ or } E = -1/2\ g\beta H \quad (4)$$

This energy-level splitting in a magnetic field is called the Zeeman effect (Fig. 7).
Interaction between the electron magnetic moment and an external magnetic field is described by the spin Hamiltonian

$$H = g\beta H S_z \quad (5)$$

The operator $\mathbf{S_z}$ acting on the α wave function of the electron gives +1/2, and it gives -1/2 with the β-function. Eigenvalues of the Hamiltonian are the quantities $+1/2g\beta\mathbf{H}$ and $-1/2g\beta\mathbf{H}$. The spin Hamiltonian formalism results in a simple description of the Zeeman effect.

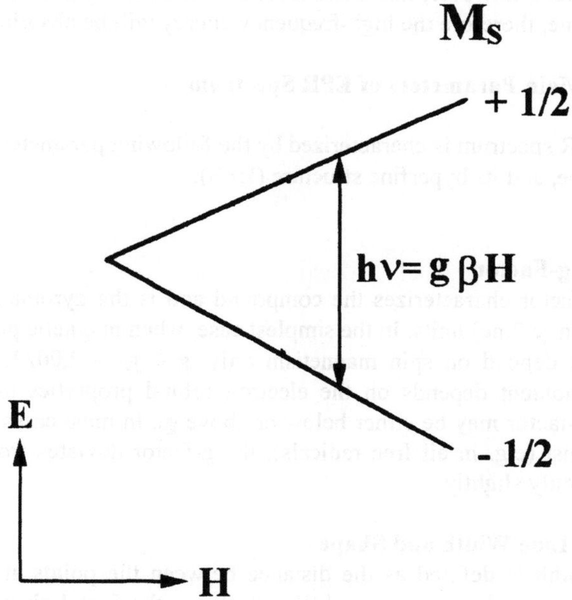

Figure 7. Zeeman levels for the free electron system.

Electron distribution between the two possible states obeys the Boltzmann law, that is,

$$n_+/n_- = \exp(-\Delta E/kT) \quad (6)$$

where n_+ and n_- are the numbers of electrons at the upper and lower levels respectively, ΔE is the difference between the energies of the levels, k is the Boltzmann constant and T is the absolute temperature.

If the frequency is

$$\nu = \Delta E/h = (E_+ - E_-)/h = g\beta H/h \quad (7)$$

then part of the electrons will leave the lower level for the upper one and absorb the alternating field energy. Simultaneously, the induced transition will occur from the upper to the lower level, leading to emission of energy.

At equilibrium, however, the lower level contains more electrons than does the upper one, therefore the high-frequency energy will be absorbed.

1.3.2 Main Parameters of EPR Spectrum

The EPR spectrum is characterized by the following parameters: g-factor, width, shape, and its hyperfine structure (HFS).

1.3.2.1 g-Factor

The g-factor characterizes the compound and is the gyromagnetic ratio expressed in (e/2mc) units. In the simplest case, when magnetic properties of the species depend on spin magnetism only, $\mathbf{g} = \mathbf{g_e} = 2{,}0023$. When the effective moment depends on the electron orbital properties as well, the effective g-factor may be either below or above $\mathbf{g_e}$. In most cases of interest for a chemist (e.g. in all free radicals), the g-factor deviates from its pure spin value only slightly.

1.3.2.2 Line Width and Shape

Line width is defined as the distance between the points at which the absorption curve slope is maximal (Fig. 8). Since the first derivative of EPR spectra is recorded, the line width is measured as the distance between extremes of the experimental $\mathbf{H_{max}}$ curve. Natural line width gives us the uncertainty in the excited state energy. The uncertainty is related by the Heisenberg relation ($\Delta E\tau \approx h$) with the mean lifetime of the species at the level considered. In frequency units, the line width is $\Delta\nu = \Delta E/h = 1/\tau$ (τ is called relaxation time).

If τ is very small, then the line will be broadened strongly and will not be observable experimentally. But if τ is too high, then the system that has absorbed the quantum will not have time to rich its Boltzman equilibrium. The electromagnetic irradiation will equalise the level populations and decrease absorption, the effect is called saturation. At a very high saturation resonance may be unobservable.

There is one more interaction type that has an effect on the width. This is called exchange interaction and is due to the fact that electrons can change their spin orientations when exchange between paramagnetic particles occurs. The exchange also averages the local interactions and, when its frequency is high, it makes the line narrower. This effect is usually observed at high concentrations of unpaired electrons in liquid or solid phases.

The line shape is the absorption intensity as a function of the magnetic field tension. If there is no exchange interaction, each of the spins is in the local magnetic field built up by a neighbouring spin; when the field obeys the Gauss distribution the line will have the Gauss shape (Fig. 8).

In liquids, the local fields are averaged to zero, so the Lorentz shape appears.

At a high concencentration of paramagnetic species in the solid phase, the exchange interaction makes the central region of the line have the Lorentz shape as well. When and where the Gauss shape transforms to the Lorentz shape depends on the exchange frequency.

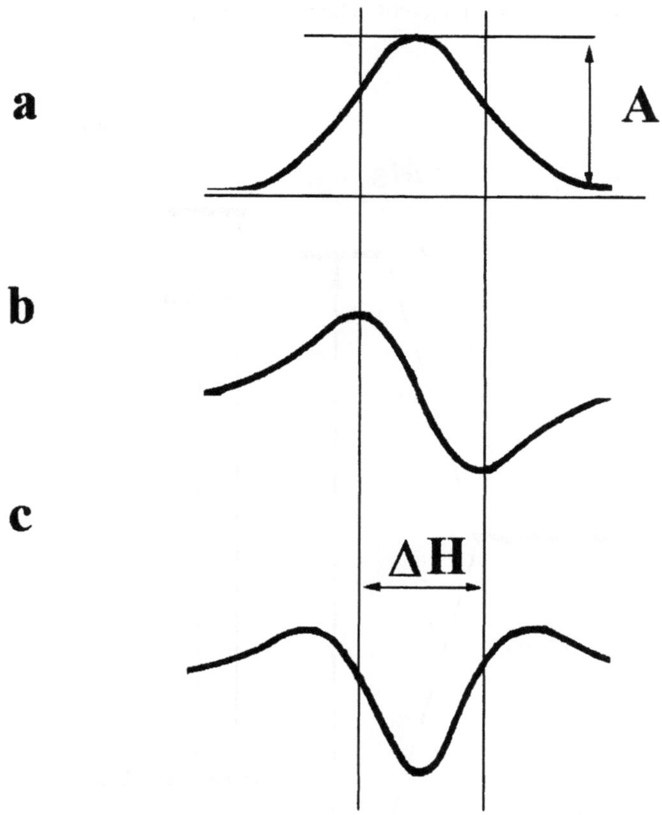

Figure 8. The absorption line shape (a) and its first (b) and second (c) derivatives.

1.3.2.3 Hyperfine Coupling Theory

When a paramagnetic species contains an unpaired electron and nuclei having their own magnetic moments (^{1}H, ^{2}D, ^{14}N, ^{13}C, ^{19}F, etc.) the coupling of electron with nuclear magnetic moments splits the singlet line into several lines. This is called hyperfine coupling while the resulting lines are called hyperfine structure (HFS) of EPR spectra.

There is a simple rule that predicts the number of HFS lines. If the nuclear spin is equal to I, the coupling will split each of the Zeeman levels into (2I + 1) sub-levels of the energies **$1/2\ g\beta H + (a/2)m_i$** . where m_i, = I, I - 1, . . . , - I, and a is the HFS constant. Figure 9 shows the systems of levels caused by coupling of a magnetic nucleus of the spin I = 1/2 with unpaired electron.

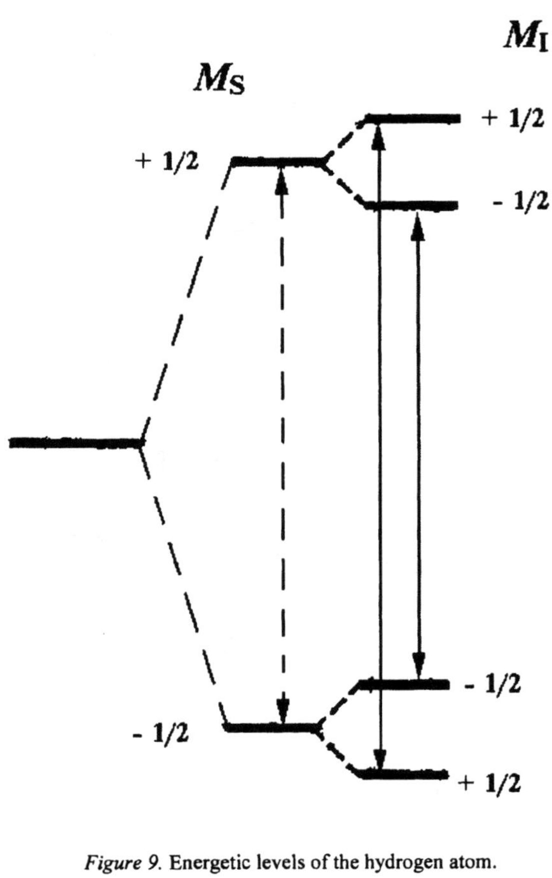

Figure 9. Energetic levels of the hydrogen atom.

Selection rules for transitions between the HFS lines are written as follows.

$$\Delta M_s = 1; \qquad \Delta m_I = 0$$

This means that orientation of nuclear magnetic moments relative to the external magnetic field does not change during the electron transitions. The selection rules provide resonance conditions for hyperfine coupling:

$$h\nu = g\beta H + am_i$$

Line intensities resulting from the coupling of an unpaired electron with one nucleus are equal, that is, all the nuclear spin projections onto the external field direction are of equal probability. The examples are atoms of hydrogen, deuterium, ions such as Mn^{+2}, and others. The spectra contain two, three, six lines for hydrogen (nuclear spin 1/2), deuterium (nuclear spin 1) and manganese (5/2) respectively (Fig. 10).

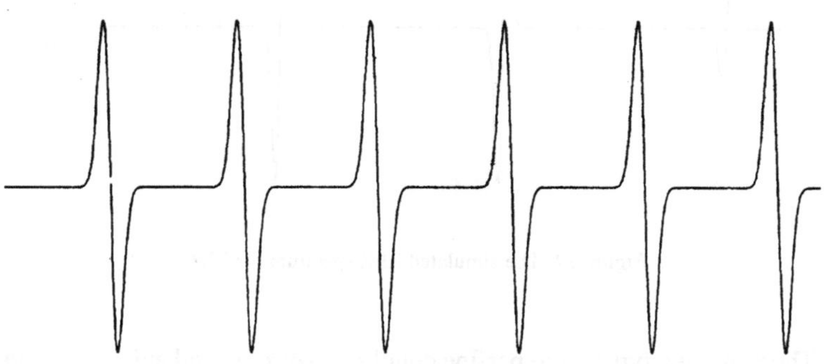

Figure 10. The simulated EPR spectrum for Mn^{+2}.

More frequent is localization of the electron near a group of atoms. This is usually observed in free radicals and radical ions whose unpaired electron is delocalized over a conjugated bonds system. Hyperfine coupling (*hfc*) in this case is more complicated.

To begin with, let hyperfine couplings with all nuclei, e.g. protons in π-electron radical, have equal energies. The respective protons are termed equivalent. The coupling with equivalent protons is described by the effective total magnetic moment and the spin of *n*/2. Consequently the number of lines is $n + 1$. As for the intensities, these are now not equal since the probabilities of the total spin projections are different, unlike the case of one nucleus. For example when an unpaired electron is coupled with two equivalent protos the possible nuclear spin orientations will correspond to the projections +1 (↑↑), 0 (↑↓) and (↓↑), and -1 (↓↓). The projection 0 is encountered twice as frequently as +1 and -1. Therefore, the central line will be twice more intense than the terminal ones (Fig. 11).

In more complex radicals different systems (referred to as A, B, C) of equivalent protons may exist. The coupling with the A protons will give (n_1+1) lines which, in turn, will be split into $(n_2 + 1)$ lines due to the B protons and, further, all the lines will be split into $(n_3 + 1)$ lines by the C protons. As a result, we have $(n_1 + 1)(n_2 + 1)(n_3 + 1)$ lines. Assignment of the complex spectra consists of finding groups of equivalent nuclei and the splitting magnitudes associated with the groups.

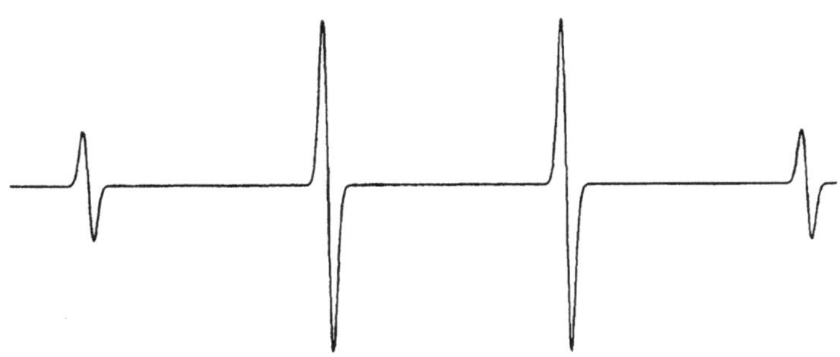

Figure 11. The simulated EPR spectrum for $^{\bullet}CH_3$

There are two types of hyperfine coupling, isotropic and anisotropic ones. Isotropic, or contact, hyperfine coupling arises when the unpaired electron density is not zero at the nucleus site. Naturally, the electron density at the nucleus is not zero only as far as S-states are concerned. Therefore, the HFS arises when the unpaired electron is either an s electron (for hydrogen or deuterium) or represented by a hybrid wave function including an s component. For the pure s electron, the HFS constant may be calculated *a priori.*

The anistropic (dipole-dipole) coupling is completely analogous to the classical interaction between magnetic dipoles. Consequently, the dipole-dipole coupling takes place only in the solid phase and preferably for paramagnetic species enclosed in a crystal.

In a study of molecular paramagnetic systems such as free radicals or complex compounds, a problem arises as to how the unpaired electron density is distributed and which is the mechanism that directs the density to an atom given. The same is valid for π-radicals in which the unpaired electron is distributed over the molecular π-orbital, a superposition of $2p_z$ atomic orbitals. On each of the carbons in a π-radical (e.g. methyl radical, triphenylmethyl radical, or various aromatic radical ions), the electron density has the symmetry inherent in the $2p_z$ orbital and has a node in the plane on which the molecule lies. In other words, to a first approximation the $2p_z$ electron interaction with a σ C–H bond should be zero. Nevertheless, EPR spectra of π-radicals do reveal hyperfine splitting due to protons.

For example, the methyl radical spectrum (Fig. 11) contains four lines, the splitting being 23.04 G and the intensities ratio 1 : 3 : 3 : 1, that is, the unpaired electron identically interacts with magnetic moments of the three hydrogens. Consequently, the unpaired electron passes ca. 23.04/506.8 = 0.041 of its time on the hydrogen 1s orbital.

This quantity is called the spin density of the unpaired electron. Rigorously speaking, the spin density is not due only to an unpaired electron, and in general, we cannot say that one electron is exactly on a given orbital while the other electrons are completely intercoupled.

The effect of an unpaired electron on other, coupled, electrons allows one to understand how the unpaired electron density leaves the $2p_z$ orbital for the hydrogen 1s orbital in the methyl radical. Usually, an isolated C–H site is considered in which the unpaired electron coupling with the σ-bond pair of electrons is studied. Two structures may be proposed (see Fig. 12)

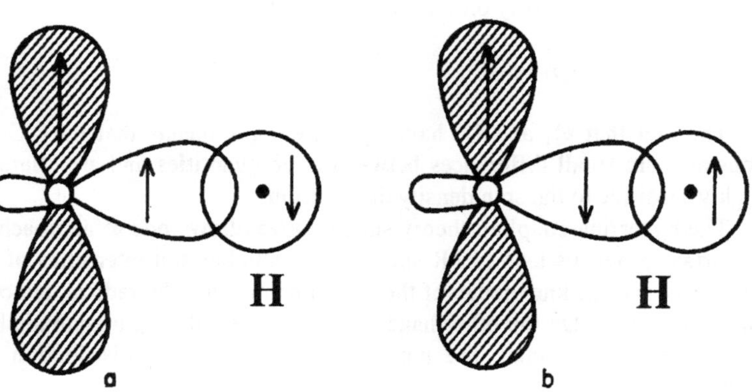

Figure 12. Hyperfine coupling of $2p_z$ electron with C–H proton.

The structure (a) is more favourable, and its weight is greater compared with the structure (b) owing to a more favourable exchange interaction between the σ- and π-electrons of the carbon whose spins are parallel. If the unpaired electron spin is α, there will be an excess of spins α on the carbon σ-orbital and the respective excess of spins β on the hydrogen 1s orbital. This is the mechanism that describes how the $2p_z$ orbital unpaired electron density can localize on the hydrogen 1s orbital and explains why the isotropic hyperfine coupling appears.

The most important corollary of the mechanism of hyperfine coupling of the carbon $2p_z$ unpaired electron with the C–H hydrogen is that the splitting a_H is proportional to the unpaired electron density on the carbon, ρ_c. This is written as follows:

$$a_H = Q_{CH}\rho_C$$

The parameter $\mathbf{Q_{CH}}$ corresponds to the splitting for which the unpaired electron density is unity. It should be equal to the splitting in the methyl radical spectrum or to the total width of the benzene radical anion spectrum, 22.5 Gauss.

The picture becomes more complicated when we compare the splitting associated with the other nuclei to the unpaired electron density distribution over the molecular system. For example, the spectra of many aromatic radicals or radical ions often reveal splitting due to the isotope ^{13}C whose natural abundance is 1%. Analysis of the splitting is often useful for estimating the densities on carbons having no hydrogens attached.

The splitting at ^{13}C depends on two parameters that interrelate it with the unpaired electron density on the three adjacent carbons:

$$a_c = Q_1\rho_1 + Q_2(\rho_2 + \rho_3)$$

The fact that $\mathbf{Q_1}$ and $\mathbf{Q_2}$ have opposite signs means that the ^{13}C HFS constants are small differences between large quantities and are, therefore, highly sensitive to the spin density distribution.

The hyperfine coupling theory stated above allows one to approach two important problems using EPR spectra. The number and intensities of EPR lines may give us knowledge of the nuclei composing the radical, hence, its chemical nature. On the other hand, the hyperfine splitting magnitude helps to qualitatively estimate the unpaired electron density distribution over atoms in the radical.

More closely one can read about EPR in the monography.[35]

1.3.3 Biradicals

Biradicals form upon binding of two molecular fragments each of which comprises one unpaired electron.[36] Exchange interaction between both paramagnetic fragments is of interest.

Upon fast exchange (i.e., upon strong interaction), the ground state of biradical can be either singlet or triplet. In the case of weak interaction, however, the EPR spectrum of biradical should be simply the combination of spectra of two independent radicals.

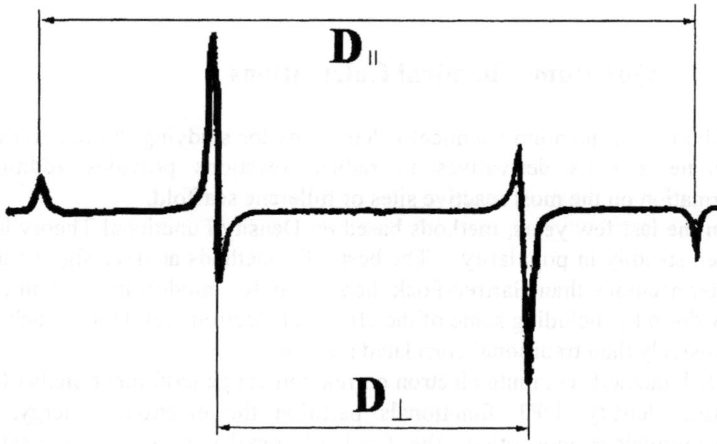

Figure 13. Theoretical EPR spectrum for randomly oriented triplet systems.

Consider a system of two magnetic moments distanced by **r** from one another. Each of the elementary dipoles builds up the local field expressed by

$$H_{loc} = g\beta S(1 - 3\cos^2\theta)/r^3$$

on the adjacent dipole site, where θ is the angle between the external magnetic field **H** direction and the vector **r**.

The most widespread case is S = 1/2. The spectrum will contain two lines, with the splitting being expressed by

$$D = g\beta(1 - 3\cos^2\theta)/r^3$$

The spin-spin coupling parameter, **D**, found experimentally may be converted into the mean distance between the paramagnetic centeres. This operation is at its simplest in the case of a single crystal where **D** varies from $\mathbf{D}_{\perp}$ to $\mathbf{D}_{\parallel} = 2\mathbf{D}_{\perp}$ in the process of the rotation. However, **D** may be found with polycrystalline samples as well. The signal can be observed in polycrystalline samples because in systems with axial symmetry the perpendicular orientation probability is much higher than that parallel, which makes the spectrum shapes so characteristic.

Figure 13 shows the spectrum calculated for a radical pair in a polycrystalline sample.

1.4 Quantum-Chemical Calculations

The use of quantum-chemical calculations for studying the reactivity of fullerene and its derivatives in radical reactions provides additional information on the most reactive sites of fullerene scaffold.

In the last few years, methods based on Density Functional Theory have gained steadily in popularity.[37] The best DFT methods achieve significantly greater accuracy than Hartree-Fock theory at only a modest increase in cost. They do so by including some of the effects of electron correlation much less expensively than traditional correlated methods.

DFT methods compute electron correlation via general functionals of the electron density. DFT functionals partition the electronic energy, the electron-nuclear interaction, the Coulomb repulsion, and an exchange-correlation term accounting for the reminder of the electron-electron interaction.

A variety of functionals have been defined, generally distinguished by the way that they treat the exchange and correlation components. Gradient-corrected functionals involve both the values of the electron spin densities and their gradients. Such functionals are also sometimes referred to as non-local in the literature (BLYP).

Although it would be desirable to calculate geometric structures of molecules with a rigorous theory, DFT methods would be computationally too expensive to use to fully optimize systems containing such big molecules like fullerenes and their derivatives. Therefore one can carry out geometry optimizations with a semiempirical method, e.g. MNDO/PM3 (AM1), and then use these geometries for DFT calculations of the spin densities.[8]

Due to the convergence problems at the UHF level, the semiempirical calculations of the fullerenyl radicals are carried out at the ROHF level.

References

1. H. W. Kroto, J. R. Heath, S. C. O'Brien, R. F. Curl, R. E. Smalley, *Nature*, 1985, **318**, 162.
2. H. B. Burgi , E. Blanc, D. Schwarzenbach, S. Liu, Y. Lu, M. M. Kappes, J. A. Ibers, *Angew. Chem., Int. Ed. Engl.*, 1992, **31**, 640.
3. R. Taylor, *Lecture Notes on Fullerene Chemistry: A Handbook for Chemists*, Imperial College Press, London, 1999.
4. R. Taylor, *J. Chem. Soc., Perkin. Trans. 2*, 1993, 813.
5. P. W. Fowler, D. E. Manolopoulos, *An Atlas of Fullerenes*, Clarendon Press, Oxford, 1995.
6. D. C. Nonhebel, J. C. Walton, *Free-Radical Chemistry*, University Press, Cambridge, 1974.
7. B. L. Tumanskii, V. V. Bashilov, S. P. Solodovnikov, V. I. Sokolov, *Bull. Russ. Acad. Sci., Div. Chem. Sci.*, 1992, **41**, 1140.
8. R. Borghi, L. Lunazzi, G. Placucci, P. J. Krusic, D. A. Dixon, N. Matsuzawa, M. Ata, *J. Amer. Chem. Soc.*, 1996, **118**, 7608.
9. P. N. Keizer, J. R. Morton, K. F. Preston, P. J. Krusic, *J. Chem. Soc., Perkin Trans. 2*, 1993, 1041.
10. I. V. Koptyug, A. G. Goloshevsky, I. S. Zavarine, N. J. Turro, P. J. Krusic, *J. Phys. Chem. A*, 2000, **104**, 5726.
11. B. L. Tumanskii, V. V. Bashilov, O. G. Kalina, V. I. Sokolov, *J. Organomet. Chem.*, 2000, **599**, 28.
12. B. L. Tumanskii, V. V. Bashilov, N. N. Bubnov, S. P. Solodovnikov, V. I. Sokolov, *Izv. Acad. Nauk, Ser. Khim.*, 1995, **1840**.
13. V. V. Bashilov, B. L. Tumanskii, P. V. Petrovskii, V. I. Sokolov, *Russ. Chem. Bull.*, 1994, **43**, 1069.
14. M. A. Cremonini, L. Lunazzi, G. Placucci, *J. Org. Chem.*, 1993, **58**, 4735.
15. J. R. Morton, K. F. Preston, P. J. Krusic, S. A. Hill, E. Wasserman, *J. Phys. Chem.*, 1992, **96**, 3576.
16. P. J. Fagan, P. J. Krusic, C. N. McEwen, J. Lazar, D. H. Parker, N. Nerron, E. Wasserman, *Science*, 1993, **262**, 404.
17. B. L. Tumanskii, E. N. Shaposhnikova, V. V. Bashilov, S. P. Solodovnikov, N. N. Bubnov, S. R. Sterlin, *Russ. Chem. Bull.*, 1997, **46**, 1174.
18. R. Borghi, L. Lunazzi, G. Placucci, G. Cerioni, A. Plumitallo, *J. Org. Chem*, 1996, **61**, 3327.
19. H. Okamura, T. Terauchi, M. Minoda, T. Fukuda, K. Komatsu, *Macromolecules*, 1997, **30**, 5279.
20. P. R. Birkett, A. G. Avent, A. D. Darwish, H. W. Kroto, R. Taylor, D. R. M. Walton, *J. Chem. Soc., Chem. Commun.*, 1993, 1230.
21. J. R. Morton, K. F. Preston, F. Negri, *Chem. Phys. Lett.*, 1994, **221**, 59.
22. P. J. Krusic, E. Wasserman, P. N. Keizer, J. R. Morton, K. F. Preston, *Science*, 1991, **254**, 1183.
23. B. L. Tumanskii, V. V. Bashilov, S. P. Solodovnikov, N. N. Bubnov, V. I. Sokolov, V. Ts. Kampel, A. Varshavskii, *Izv. Acad. Nauk, Ser. Khim.*, 1994, 673.
24. P. N. Keizer, J. R. Morton and K. F. Preston, *J. Chem. Soc., Chem. Commun.*, 1992, 1259.
25. R. G. Gasanov, O. G. Kalina, V. V. Bashilov, B. L. Tumanskii, *Russ. Chem. Bull.*, 1999, **48**, 2344.
26. M. Yoshida, F. Sultana, N. Uchiyama, T. Yamada, M. Iyoda, *Tetr. Lett.*, 1999, 735.

27. B. L. Tumanskii, O. G. Kalina, V. V. Bashilov, A. V. Usatov, E. A. Shilova, Yu. I. Lyakhovetsky, S. P. Solodovnikov, N. N. Bubnov, Yu. N. Novikov, A. S. Lobach, V. I. Sokolov, *Russ. Chem. Bull.*, 1999, **48**, 1108.
28. B. L. Tumanskii, O. G. Kalina, V. V. Bashilov, *J. Mol. Mat.*, 2000, **13**, 23.
29. R. Klemt, E. Roduner, H. Fischer, *Chem. Phys. Lett.*, 1994, **229**, 524.
30. T. Tanaka, K. Komatsu, *J. Chem. Soc., Perkin Trans. 1*, 1999, 1671.
31. P. J. Fagan, P. J. Krusic, D. H. Evans, S. A. Lerke, E. Johnston, *J. Am. Chem. Soc.*, 1992, **114**, 9697.
32. P. Zhou, G.-Q. Chen, H. Hong, F.-S. Du, Z.-C. Li, F.-M. Li, *Macromolecules*, 2000, **33**, 1948.
33. R. Gregory, R. N. Haszeldine, A. E. Tipping, *J. Chem. Soc. C*, 1970, 1750.
34. N. N. Bubnov, S. P. Solodovnikov, in *Modern Physics in Chemistry*, Academic Press, London, 1976, 53.
35. J. E. Wertz, J. R. Bolton, *Electron Spin Resonance*, McGraw-Hill Book Company, New York, 1972.
36. S. H. Glarum, J. H. Marshal, *J. Chem. Phys.*, 1967, **47**, 1374.
37. J. B. Foresman, A. Frisch, *Exploring Chemistry with Electronic Structure Methods*, Gaussian, Inc., Pittsburgh, 1996.

Chapter 2

Radical Addition to C_{60}

A great number of weakly conjugated double bonds in fullerenes make them reactive toward different radicals. By now, the structures and reactivities of adducts of fullerene-60 with carbon-, oxygen-, sulfur-, phosphorus-, boron-, silicon, and metal-centered radicals, as well as with hydrogen and fluorine atoms have been studied by EPR.[1-5] It is possible to isolate new paramagnetic derivatives of C_{60} using improved HPLC separations.

The structure of radical adducts of free radicals with fullerenes and their reactivity depend on the number of the addends added. First of all, it was very important to answer the questions:

1. How is the unpaired electron delocalized in these radicals?
2. What are the main paths of their decay?
3. What are their spectral characteristics?

The main features of the fullerenyl radicals were revealed for adducts of the fullerene with organic and organoelement free radicals. This is why we will begin this chapter with the description of the reactions of these radicals with C_{60}. Then, the features of the addition of atoms to fullerenes will be described.

1.1 Addition of Carbon-Centered Radicals

The addition of alkyl radicals to C_{60} was the first well studied radical reaction of fullerenes.[6-8] EPR studies of resultant radical adducts made it possible to determine the character of delocalization of an unpaired electron in fullerenyl radicals and their reactivity.

Hfc constants for alkyl radical adducts are listed in Table 1. The individual lines in the spectra of $^{\bullet}C_{60}Alk$ are narrow (0.02 G).[9] The formation of monoadducts was confirmed by analysis of the proton and ^{13}C (where possible) hyperfine structures.

Table 1. *Hfc* constants and dimerization enthalpies for alkyl fullerenyl radicals.

$R^•$	$a_{H(D)}$, G	a_{13C}, G	ΔH, kcal mol^{-1}	T, K	Ref.
CH_3	3 H = 0.035			300	9
$MeCH_2$	2 H = 0.28			425	9
	3 H = 0.12				
$CH_2(OH)$	2 H = 0.175			300-	18
	1 H = 0.085			400	
CH_2CO_2H	2 H = 0.33			300-	18
				400	
$Me^{13}CH_2$		1 C = 15.5		473	8
Me_2CH	1 H = 0.47		35.5	440	8,9
	6 H = 0.14				
Me_3C	9 H = 0.17	1 C = 17.8	22.0	325	8,9
		1 C = 13.1			
		2 C = 9.39			
		2 C = 8.86			
		3 C = 5.59			
		2 C = 4.48			
		2 C = 4.02			
$Me_3{}^{13}C$	9 H = 0.17	1 C = 13.1		370	8
$(CD_3)_3C$	9 D = 0.028	2 C = 3.59		370	8
		4 C = 2.41			
		8 C = 0.84			
$(MeCH_2)_3C$	3 H = 0.34	1 C = 0.40		370	8
	3 H = 0.17				
$C_{10}H_{15}$	3 H = 0.25	1 C = 17.75	21.6	300-	8
	6 H = 0.044	1 C = 12.34		400	
		2 C = 9.30			
		2 C = 8.79			
		3 C = 5.59			
		2 C = 4.48			
		2 C = 4.03			
CCl_3		1 C = 29.6	17.1	250-	8
		1C = 18.1		310	
		4 C = 8.8			
CBr_3		1 C = 35.3	17.0	300-	8
		1 C = 18.0		380	
$PhCH_2$	2 H = 0.42			350	8
	2 H = 0.19				
$Ph^{13}CH_2$	2 H = 0.42	1 C = 14.9		350	8
	2 H = 0.19				
C_6H_5	2 H = 0.22				11
m-MePh	1 H = 0.23				11
p- MePh	2 H = 0.22				11
1-naphthyl	2 H = 0.25				11

$R^{\bullet}$	$a_{H(D)}$, G	a_{13C}, G	ΔH, kcal mol^{-1}	T, K	Ref.
2-naphthyl	1 H = 0.18				11
9-phenanthryl	1 H = 0.26				11

The most intense and informative EPR spectra were recorded for the radical adducts of C_{60} with a *tert*-butyl radical and its deuterated analog.[7] Radical adduct $^{\bullet}C_{60}CMe_3$ was obtained by UV irradiation (T = 323 K) of a saturated benzene solution of C_{60} containing $BrCMe_3$. Its analog $^{\bullet}C_{60}C(CD_3)_3$ was prepared by photolysis of di-*tert*-butyl peroxide in a benzene solution containing C_{60} and deuterated isobutane.

The EPR spectrum of the $^{\bullet}C_{60}C(CD_3)_3$ radical exhibits ten pairs of satellites, which were assigned to the interaction of the unpaired electron with ^{13}C (I=1/2) (Fig. 1) of the fullerene core.

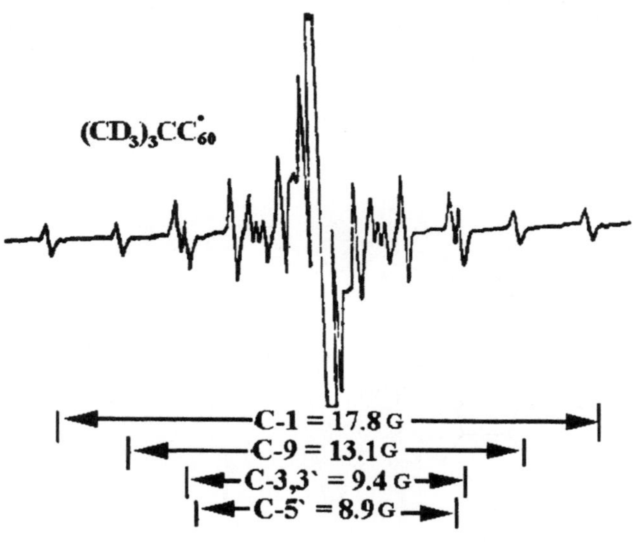

Figure 1. EPR spectrum of $^{\bullet}C_{60}C(CD_3)_3$ showing ^{13}C satellites. Adapted from ref 7.

Their relative intensities and *hfc* values indicate that the radical adduct has C_s symmetry and that the unpaired electron is localized in six-membered rings, adjacent to the C-CR bond.[7]

The isotropic *hfc* with ^{13}C nuclei characterizes the unpaired electron density in the s orbitals of carbon atoms and provides no information on its density in the 2p orbital. Based on assumption that *hfc* with the C(1), C(3), C(3'), C(5) and C(5') atoms are proportional to their spin density in 2p orbitals, the spin density for the C(1) atom was found to be 0.33, and those for the C(3), C(3'), C(5), and C(5') were found to be 0.17.

A similar pattern of the *hfc* of an unpaired electron with ^{13}C nuclei was observed for other radical adducts of alkyl and phosphoryl radicals with C_{60}.[7,10]

Semiempirical and DFT calculations agree well with the conclusion that atoms C(1), C(3), and C(3') show large positive *hfc*, whereas C(2) and C(2') have large, negative hyperfine interactions.[4] It appears that spin-polarization effects are important:

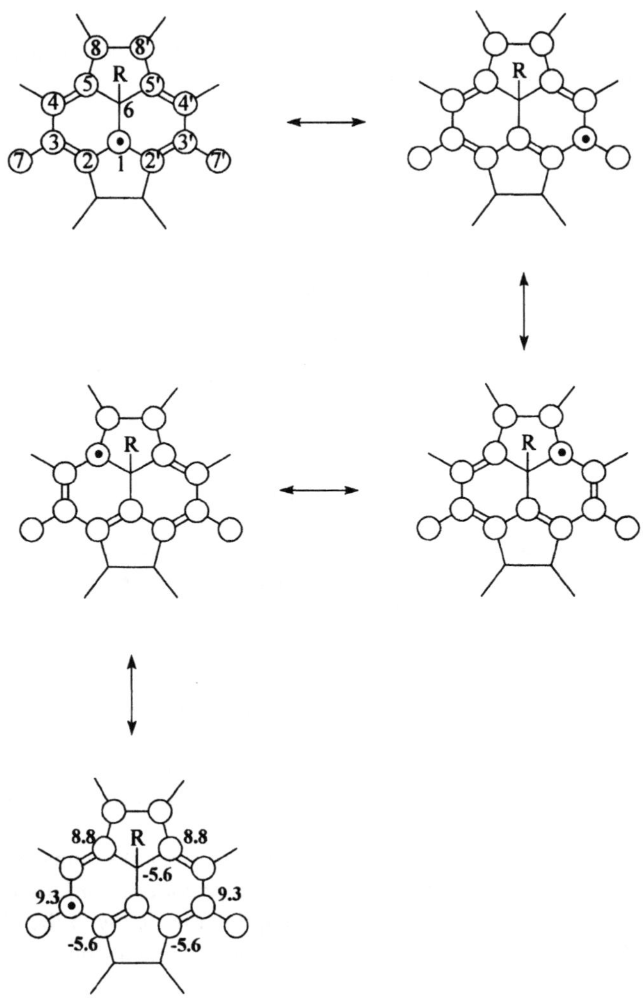

Unusual proton hyperfine structures were observed for benzyl and 1-adamantyl adducts.[7,8] Both the radicals were generated from the corresponding hydrocarbons by hydrogen abstraction with *tert*-butoxyl radical.

The benzyl adduct showed a triplet of triplets (two benzylic protons and only the two *ortho* protons of the phenyl ring). In 1-adamantyl-$C_{60}^{\bullet}$ the six protons nearest to the C_{60} surface had smaller hyperfine interactions than the three more remote *tert* protons.

Intense UV photolysis of dilute benzene solutions containing C_{60} and C_6H_5–X (X = Br, I or HgC_6H_5) produces radical adduct $^{\bullet}C_{60}C_6H_5$ (Table 1) whose EPR spectrum is a 1 : 2 : 1 triplet (0.22 G) due to the hyperfine interaction of the unpaired electron with the pair of phenyl protons in the *meta* position.[11] This assignment is established by the spectral changes that are observed when these protons are replaced by various substituents. Thus, addition of the aryl radicals obtained from 3-methylbromobenzene and 3,5-dimethylbromobenzene to C_{60} yields adducts whose spectra display a doublet and a singlet since only one hydrogen or none are left in the *meta* position, respectively.[11]

A binomial triplet is observed for radical obtained by photolysis of C_{60} and 4-methylbromobenzene as both *meta* protons are present. A relatively strong hyperfine interaction in the *meta* position and no measurable interaction in the *ortho* and *para* positions is also revealed for fluorophenyl analogs. Thus, 3,5-difluorobromobenzene reacts with C_{60} to give radical adduct displaying only a hyperfine interaction of 0.60 G for two equivalent fluorines, while 3-fluorobromobenzene yields radical with a doublet of doublets pattern for one *meta* fluorine (0.50 G) and one *meta* proton (0.29 G). A slightly larger F hyperfine interaction compared with the proton interaction in the same position in aromatic radicals is well established.[11]

No fluorine hyperfine splitting is detected for radical derived from 4-fluorobromobenzene, since fluorine occupies the *para* position, and only a splitting due to two equivalent *meta* protons is observed.[11]

EPR spectra of the products of photoreaction of C_{60} with 1- and 2-bromonaphthalene agree with the above behavior of hyperfine splittings. For example, in C_{60}-1-naphthyl radical adduct, the substituent bears two *meta*-like protons and thus displays a triplet spectrum. The similarity is not complete at much higher resolution as evidenced by a broadened central line of the 1 : 2 : 1 triplet, and two additional smaller splittings (0.07 G and 0.03 G) each due to single protons can be discerned. In the radical adduct with 2-naphthyl substituent, on the other hand, the substituent has only a single *meta* proton that accounts for the observed doublet spectrum (Fig. 2). Similarly, the photoreaction of C_{60} with 9-bromophenanthrene yields radical, which displays a doublet spectrum since there is only one proton of the 9-phenanthryl moiety that occupies a *meta*-like position.[11]

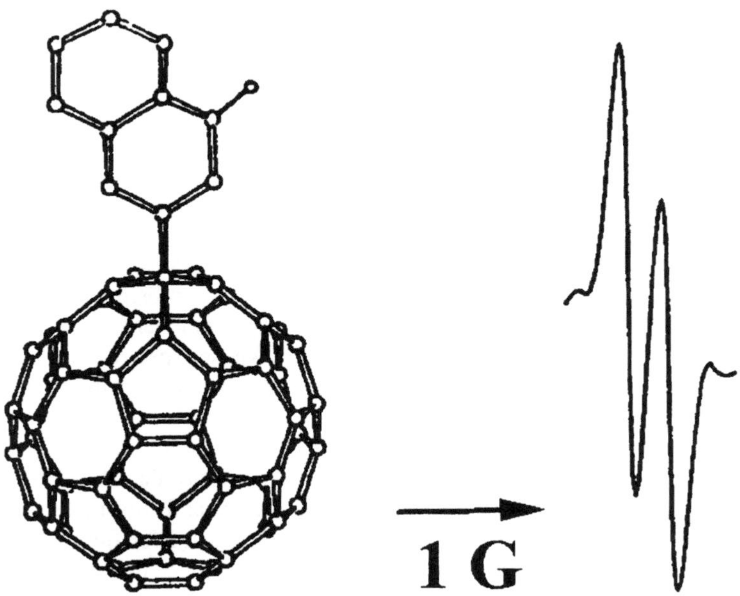

Figure 2. EPR spectrum of 2-naphthyl-$C_{60}^{\bullet}$. Adapted from ref 11.

Unusual long-range interactions were observed for the adamantyl and bicyclooctyl radical adducts of C_{60} suggesting that effects similar to those described by the "W plan-in" long-range hyperfine interactions are operative.[7,12]

Photolysis of solutions containing HgR_2 (R = CH_2CH_3, $CH(CH_3)_2$, CH_2Ph) in the presence of [60]fullerene have been investigated by transform time-resolved EPR (FT TR EPR) and continuos-wave EPR (CW EPR) techniques.[13]

Both electron-spin-polarized $^3C_{60}$ (*A* polarization) and electron-spin-polarized adducts $^{\bullet}C_{60}R$ were observed by FT TR EPR. The CW EPR spectra of the $^{\bullet}C_{60}R$ under steady-state irradiation also exhibit certain electron-spin polarization.

The chemically induced dynamic electron polarization (CIDEP) in the FT TR EPR experiments is explained by the following series of steps. Initially, photolysis causes the cleavage of the HgR_2 into radicals combining with C_{60} to give $^{•}C_{60}R$, which dimerize to form $RC_{60}–C_{60}R$. In the FT TR EPR experiments, laser excitation of C_{60} produces $^3C_{60}$ and photocleavage of the dimers results in $^{•}C_{60}R$.

The observed *E*/*A* CIDEP patterns at short (< 1 μs) delays after the laser flash are supposed to be a result of polarization through the radical-pair mechanism resulting from the interactions of two $^{•}C_{60}R$ radicals formed via the photocleavage of $RC_{60}–C_{60}R$.

The additional *E* polarization observed at later times (>1 μs) is assumed to result from the interaction of $^3C_{60}$ with $^{•}C_{60}R$ to cause *E* polarization through the radical-pair-triplet mechanizm (Fig. 3).

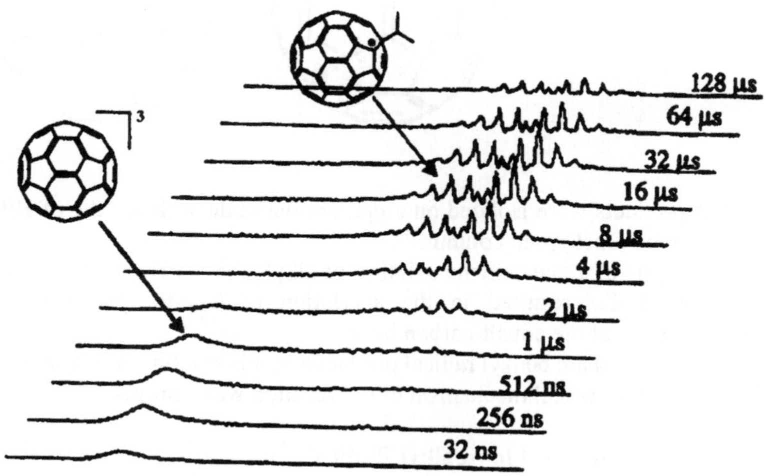

Figure 3. Evolution of the FT TR EPR intensity for $^{•}C_{60}Pr^{i}$. Adapted from ref 13.

The polarization observed in the CW EPR experiments is attributed to the maintenance of polarization during the radical lifetime because of extremely long spin-lattice relaxation of the $^{•}C_{60}R$.[13]

Thermal decomposition of azo-*bis*-(isobutyronitrile) in a 1,2-dichlorobenzene solution of C_{60} gave three regioisomeric fullerene derivatives, two of which were assigned as the 1,4- and 1,2-regioisomers:[14]

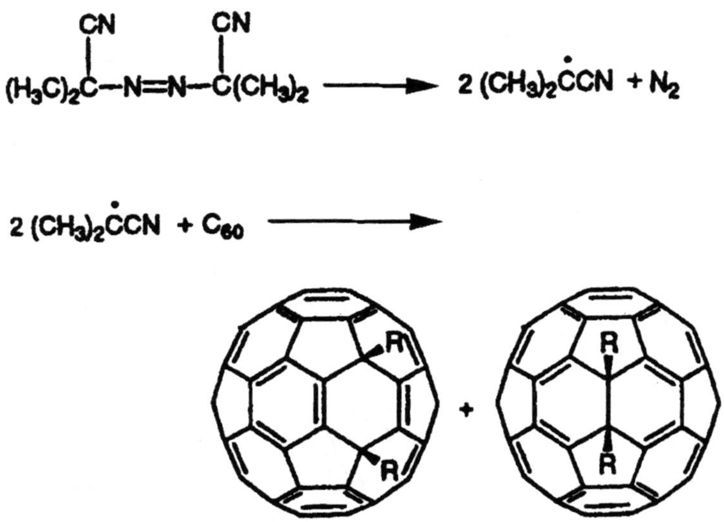

The products were isolated on a preparative scale with the use of HPLC on a Cosmoil Buckyprep column.[14]

Photolysis of a benzonitrile solution of alkylcobalt (III) complexes in the presence of C_{60} resulted in the alkylation of C_{60} to give R_2C_{60} via photocleavage of the cobalt-carbon bond.

The intermediate benzyl radical produced in this reaction was detected by EPR. Thus, the radical mechanism of the reaction was confirmed.[15]

1.1.1 Addition of Fluoroalkyl Radicals

Besides many alkyl-$C_{60}^{\bullet}$ radicals, fluoroalkyl-C_{60} radicals have been also detected (Table 2).[16-22]

Table 2. Hfc constants for flouroalkyl fullerenyl radicals.

R	a_H, G	a_F, G	T, K	Ref
CH_2F	0.23	0.78	295	19
CHF_2	0.15	0.24	295	19
CH_3CF_2	0.08	0.15	295	19
CF_3CH_2	0.42	2.72	295	19
CF_3CHF	0.53	3 F = 2.77 1 F = 0.65	295	19
$CF_3(CH_3)CH$	1 H = 0.55 3 H = 0.14	2.62	295	19
$CF_3(CH_3)CH$	1 H = 0.55 3 H = 0.14	3 F = 2.62	250-425	18

R	a_H, G	a_F, G	T, K	Ref
$(CF_3)_2CH$	0.52	2.01	295	19
CF_3		1F = 0.63, 2F = 0.28	150-200	17
		3F = 0.074		
		3F = 0.18	285	
			425	
CF_3CF_2		2F = 0.33, 3F = 2.43	225	17
		2F = 0.53, 3F = 2.30		
			450	
$(CF_3)_2CF$		1F = 0.33,	150-200	17
		1F = 0.83, 6F = 2.00,	300	
		1F = 1.02, 6F = 2.05		
			450	
3,5-difluorophenyl		2 F = 0.60		11
3-fluorophenyl	1 H = 0.29	1 F = 0.50		11
4-fluorophenyl	2 H = 0.20			11
$(CF_3)_3C$		1F = 2.92, 2F = 1.93,	225	17
		2F = 4.49, 0.27, 0.13		

A convenient method for the preparation of fluoroalkyl-modified C_{60} using fluoroalkyl halides and Bu_3SnH under radical conditions has been developed.[20]

Preparation of fluoroalkyl-modified C_{60}. *A benzene solution of $BrCF_2CO_2Et$ (495 mg, 2.4 mmol), C_{60} (144 mg, 0.2 mmol), and Bu_3SnH (292 mg, 1.0 mmol) was refluxed under nitrogen for 30 h. The products were separated by gel permeation chromatography using toluene as eluent to give 30 mg of the product (37% based on consumed C_{60}).*[20]

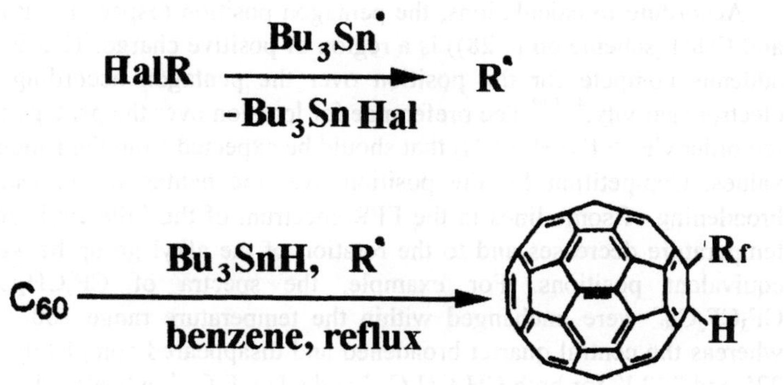

$R = CF_2CO_2Et, C_6F_{13}, CF_2Br, C_{12}F_{25}, (CF_2)_6I$

An interesting feature of C_{60}-fluoroalkyl radicals is connected with their structure, which follows from the comparison of the spectra of $CH_3CH_2C_{60}^{\bullet}$, $CH_3CF_2C_{60}^{\bullet}$, $CF_3CH_2C_{60}^{\bullet}$, and $CF_3CF_2C_{60}^{\bullet}$ (Fig. 4).[4]

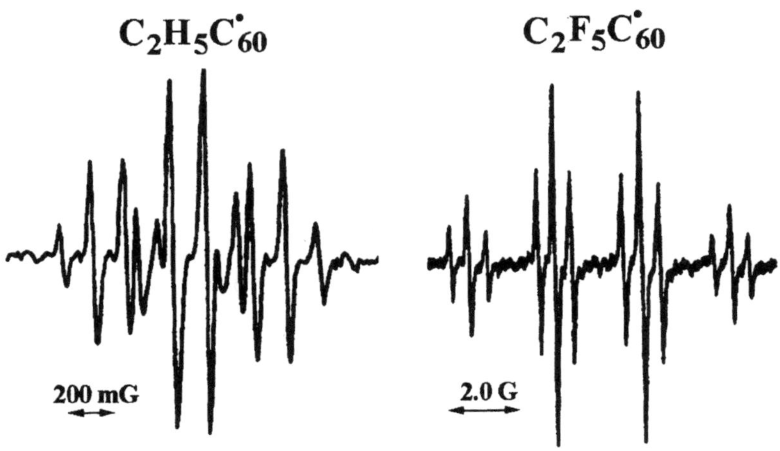

Figure 4. EPR spectra of $C_2H_5C_{60}^{\bullet}$ and $C_2F_5C_{60}^{\bullet}$ at high temperatures. Adapted from ref 4.

The value and the sign of the hyperfine interaction of unpaired electron with H and F nuclei has been found to be affected by the position of the addend relative to the C_s symmetry axis of the fullerenyl radical, which depends on the electrostatic forces between the ligands and C_{60} surface.

According to calculations, the pentagon position (especially atoms C(8) and C(8') (scheme on p. 28)) is a region of positive charge. That is why the addends compete for the position over the pentagon according to their electronegativity.[4,18,19] The preference for location over the pentagon follows the order $CF_3 > F > H > CH_3$ that should be expected from their inductive σ-values. Competition for the position over the pentagon can lead to the broadening of some lines in the EPR spectrum of the fullerenyl radicals as temperature decreases and to the rotation of the alkyl group between non-equivalent positions. For example, the spectra of $CF_3CH_2C_{60}^{\bullet}$ and $CF_3CF_2C_{60}^{\bullet}$ were unchanged within the temperature range 160 – 450 K, whereas the central quartet broadened and disappeared completely between 225 and 240 K for both $CH_3CH_2C_{60}^{\bullet}$ and $CH_3CF_2C_{60}^{\bullet}$ radicals. The matter is that the equilibrium conformations of $CF_3CH_2C_{60}^{\bullet}$ and $CF_3CF_2C_{60}^{\bullet}$ are symmetric with respect to the CF_3 group over the pentagon, whereas the equilibrium conformations of $CH_3CH_2C_{60}^{\bullet}$, $CH_3CF_2C_{60}^{\bullet}$ are asymmetric (the CH_3 group is over one of the hexagons):[4,18,19]

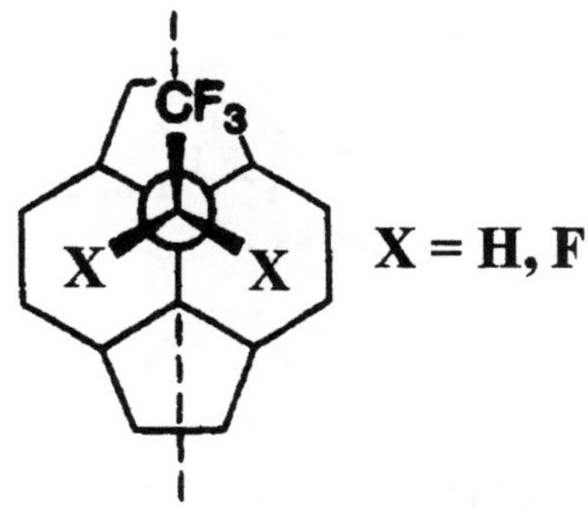

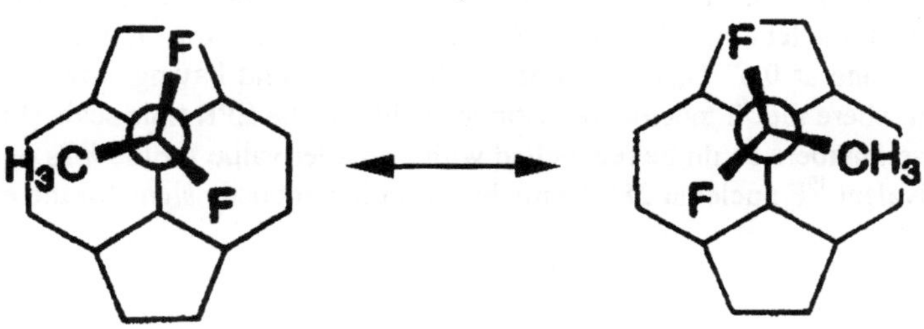

From the analysis of the values of the hyperfine interaction of unpaired electron with H and F nuclei in methyl and trifluoromethyl group, one can determine their position over the pentagon or hexagon.

For example,[9] in $(CH_3)_3C_{60}^{\bullet}$ at 225 K, the nine proton hyperfine interactions consist of two groups: three at 0.34 G (over pentagon) and six at 0.088 G (over the hexagons).

In $CH_3CF_2C_{60}^{\bullet}$, the *hfc* with the methyl group, which is also over one of the hexagons, is 0.080 G. As usual, the observed range of the *hfc* values for CH_3 over hexagon is 0.08 - 0.140 G.

The *hfc* values for the trifluoromethyl group over pentagon in the polyfluoroalkyl-C_{60} radicals equal to 2.72 G in $CF_3CH_2C_{60}^{\bullet}$, 2.43 G in $CF_3CF_2C_{60}^{\bullet}$, 2.77 G in $CF_3CHFC_{60}^{\bullet}$, and 2.27 G in $CF_3(CH_3)CHC_{60}$ radical.

If there is no another possibility, CF_3 group occupies the position over pentagon. Only for $(CF_3)_3C_{60}^{\bullet}$ at 225 K, it is possible to place two CF_3 groups over the hexagons (1.63 G) and one over the pentagon (2.26 G).

There is a strong evidence for $CF_3C_{60}^{\bullet}$ that a ^{19}F nucleus over the pentagon ($\theta = 0°$) has *hfc* which is opposite in sign to those over the hexagons ($\theta = 120°$):[17,19]

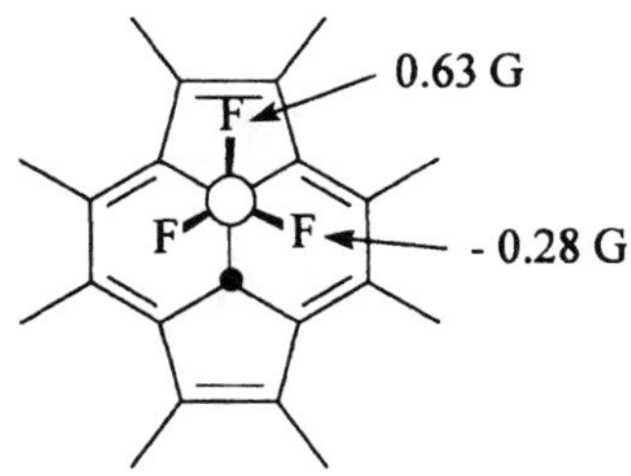

At 285 K, the EPR spectrum of $CF_3C_{60}^{\bullet}$ shows hyperfine interaction with three ^{19}F nuclei ($a_F = 0.074$ G). At 180 K it is that of two ^{19}F nuclei at 0.28 G and one at 0.63 G, rotation about the C–C_{60} bond having slowed to the point where the ^{19}F nuclei are distinguishable on the EPR time scale (Fig. 5). These numbers could be reconciled with a smaller value for the *hfi*'s of three equivalent ^{19}F nuclei at 285K only by assuming opposite signs for the *hfi*'s at 180 K.

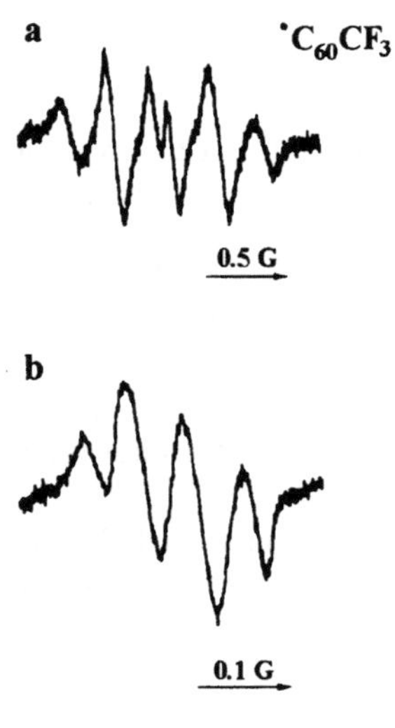

Figure 5. EPR spectrum of $C_{60}CF_3$ at (a) 185 K and (b) 285 K. Adapted from ref 17.

Calculation on the CH_3-containing fragment of C_{60}, where it represents a structure corresponding to approximately one-half of the C_{60}, confirm positive and negative spin populations at $\theta = 0°$ and $120°$, respectively.

Radical adducts of branched perfluorinated σ-radicals with C_{60} are characterized by hyperfine interaction with the nucleus of the δ-fluorine atom located above the pentagon in fullerenyl radical.[21]

The 3,3-*bis*(trifluoromethyl)perfluorobut-1-enyl radical was generated by photolysis of the parent mercury compound in a saturated solution of C_{60} in 1,2,4-trichlorobenzene in the temperature range from 270 to 430 K. The perfluoroisobutyryl radical was synthesized by UV photolysis of the branched α-triketone:

$$Hg[CF{=}CFC(CF_3)_3]_2 \xrightarrow{h\nu} Hg + 2\,{}^{\bullet}CF{=}CFC(CF_3)_3$$

$$C_{60} + {}^{\bullet}CF{=}CFC(CF_3)_3 \longrightarrow (CF_3)_3CFC{=}CF{-}C_{60}{}^{\bullet}$$

$$(CF_3)_2CFC(CO)(CO)(CO)CCF(CF_3)_2 \xrightarrow{h\nu} 2\,{}^{\bullet}C(CO)CF(CF_3)_2 + CO$$

$$C_{60} + {}^{\bullet}C(CO)CF(CF_3)_2 \longrightarrow (CF_3)_2CF(CO)C{-}C_{60}{}^{\bullet}$$

By analogy with other adducts of fluoroorganic radicals with C_{60}, the *hfi* with the nucleus of δ-F atom located above the pentagon in the radical $(CF_3)_3CF_{\delta}C{=}CF_{\gamma}\text{-}C_{60}$ will be larger than the *hfi* with the nucleus of the γ-F atom, and hence *hfi* constant 0.35 G is caused by the γ-F atom (Fig. 6). In the radical $(CF_3)_2CF_{\delta}(CO)C\text{-}C_{60}$, the observed splitting of the doublet (0.2 G) is associated with the interaction of the unpaired electron with the nucleus of the δ-F atom. Heating of these samples to 460 K results in line broadening due to an increase in the rate of exchange between non-equivalent positions:[21]

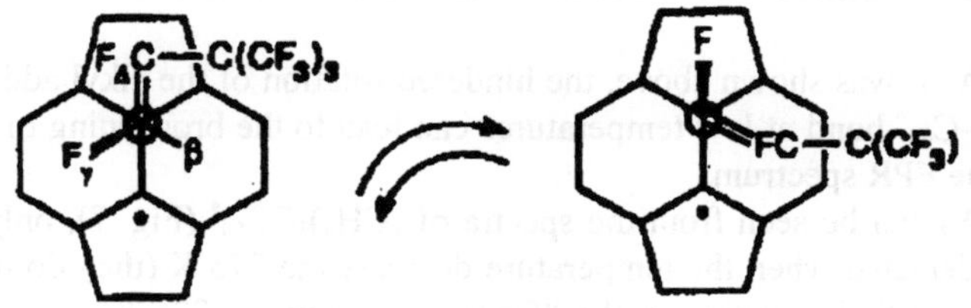

A *B*

The fact that the broadening of the doublet is observed at higher temperatures than that for alkyl radicals is connected with relatively high rotation barrier, which may be associated with the interaction between the unpaired electron and the double bond.

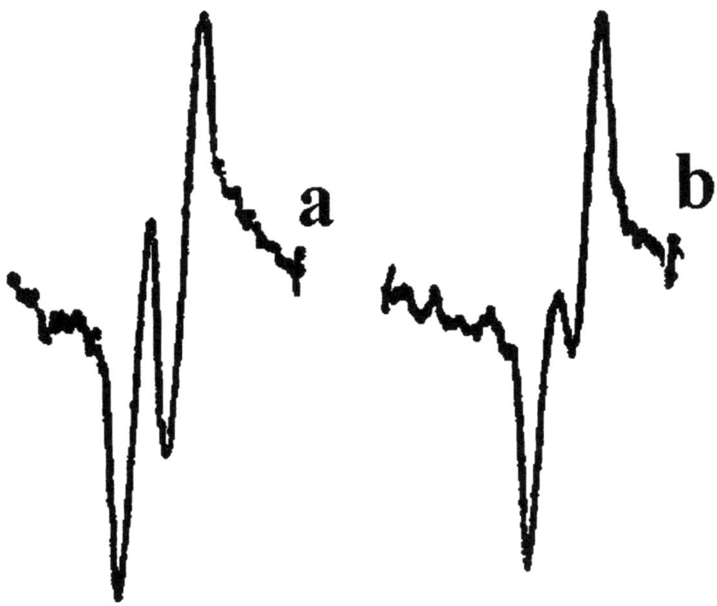

Figure 6. EPR spectra of $(CF_3)_3CFC{=}CF\text{-}C_{60}^{\bullet}$ at (a) 270 K and (b) 400 K.[21]

1.1.2 Hindered Rotation in Alkylfullerenyl Radicals

As it was shown above, the hindered rotation of the alkyl addend around Alk–$C_{60}^{\bullet}$ bond at low temperatures can lead to the broadening of some lines in the EPR spectrum.

As can be seen from the spectra of $(CH_3)_3CC_{60}^{\bullet}$ (Fig. 7), only two lines are detected when the temperature decreases to 275 K (they do not broaden because their position in the "frozen" spectrum at 225 K is the same as at room temperature).[9]

Analysis of broadening of lines of $(CH_3)_3CC_{60}^{\bullet}$ allowed one to estimate the barrier of the hindered rotation: ΔH = 7.3 kcal mol^{-1}. The height of the energy barrier is comparable with the barriers to the hindered rotation of the *tert*-butyl group in substituted ethanes (9.6 – 10 kcal mol^{-1}) and in $^{t}BuC_{60}^{-}$ anion (9.3 kcal mol^{-1}).[9,23]

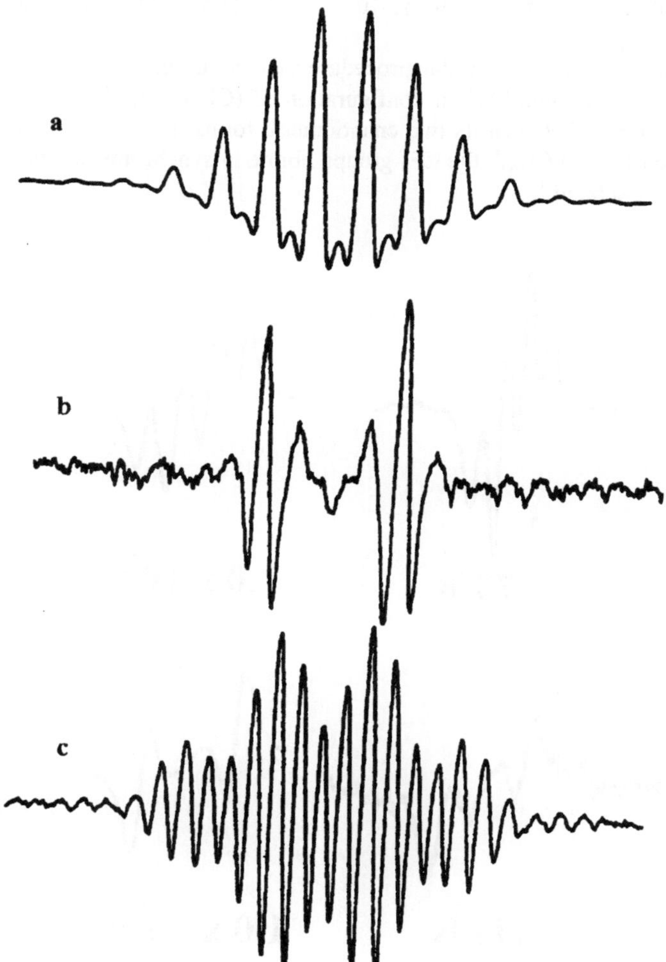

Figure 7. Second derivative EPR spectra of $^{\bullet}C_{60}Bu^{t}$ at (a) 325 K, (b) 275 K, (c) 225 K. Adapted from ref 9.

The effects of line broadening in $CH_3C_{60}^{\bullet}$ radical in the temperature range 200 – 275 K has been simulated using the Bloch equations modified for chemical exchange, which reproduce the changes in the spectrum remarkably well (Fig. 8).[9]

The rate constants for the rotation about CH_3–C_{60} bond and the barrier to the rotation were obtained to be 10 – 100 MHz and ΔH = 3.3 kcal mol^{-1}, respectively.[9]

In case of CF_3, the similar procedure yielded an enthalpy barrier of 7.0 kcal mol^{-1}.[9] The equilibrium configuration of $(CF_3)_2CFC_{60}^{\bullet}$ is asymmetric, and the barrier between its two enantiomeric forms is *ca.* 3.5 kcal mol^{-1}. In the case of $(CF_3)_3CC_{60}^{\bullet}$, the CF_3 groups librate in synchronism with rotation about C–C_{60} bond.[17]

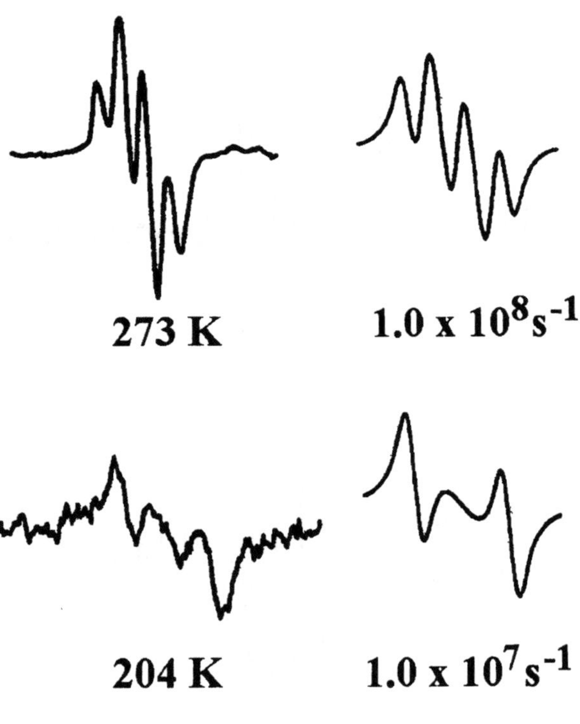

Figure 8. EPR spectra of $^{\bullet}C_{60}CH_3$ at different temperatures and their simulations. Adapted from ref 9.

The temperature-dependent broadening of the central line in the EPR spectra of $CF_3(CF_2)_n$-$C_{60}^{\bullet}$ was observed for perfluoro-*n*-alkyl adducts of C_{60}. It is also attributed to the hindered rotation of the perfluoroalkyl radical between two non-equivalent positions (Fig. 9).[16]

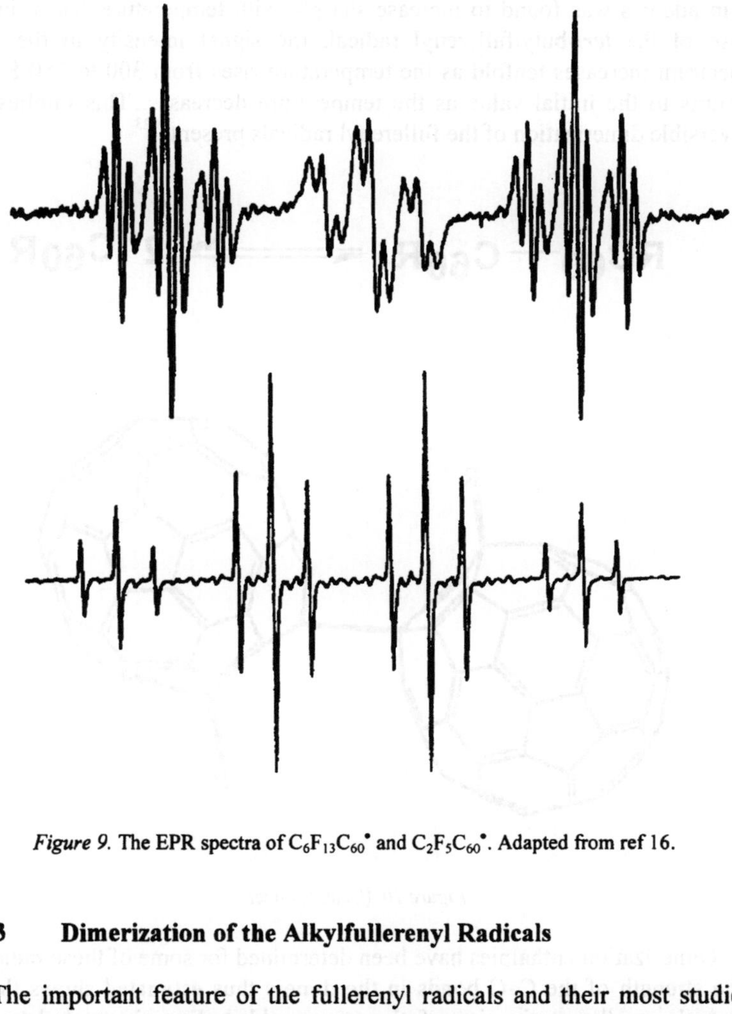

Figure 9. The EPR spectra of $C_6F_{13}C_{60}^{\bullet}$ and $C_2F_5C_{60}^{\bullet}$. Adapted from ref 16.

1.1.3 Dimerization of the Alkylfullerenyl Radicals

The important feature of the fullerenyl radicals and their most studied reaction is the reversible dimerization of the head-to-head type (Fig. 10).[1,2,8,16,22,23]

The major portion of the fullerenyl radicals is destroyed after irradiation ceased. However, the intensity of the EPR signals of some radicals does not decrease when the irradiation is stopped, and the EPR signal can be detected for several hours or even days. This behaviour is essentially typical for radicals containing bulky fragments, namely, *tert*-butyl, isopropyl, or trichloromethyl groups. In addition, the intensity of spectral signals of these spin adducts was found to increase sharply with temperature. Thus, in the case of the *tert*-butylfullerenyl radical, the signal intensity in the EPR spectrum increases tenfold as the temperature rises from 300 to 350 K and returns to the initial value as the temperature decreases. This implies the reversible dimerization of the fullerenyl radicals present.[8,23]

$$RC_{60}\text{—}C_{60}R \rightleftharpoons 2\ {}^{\bullet}C_{60}R$$

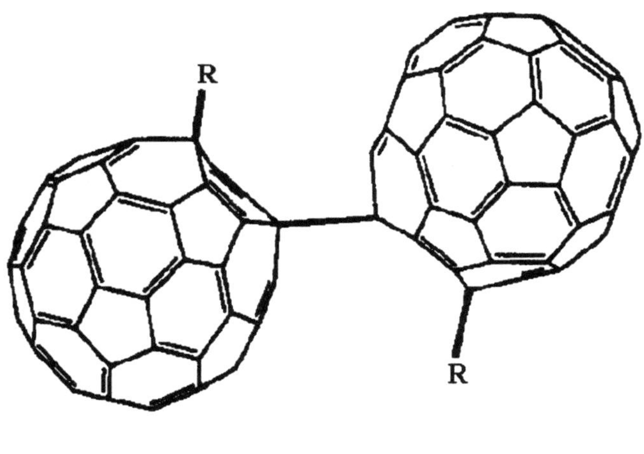

Figure 10. $(C_{60}R_2)_2$ dimer.

Dimerization enthalpies have been determined for some of these radicals. The strength of the C–C bonds in the dimers thus estimated shows that it depends on the steric size of the group added. Experimental data and molecular mechanical calculations showed that the dimerization of fullerenyl radicals follows the head-to-head pattern and involves C(3) and C(3') atoms.

Steric restrictions prevent the dimerization through the cage C(1) atoms, which carry high electron density.[8]

An efficient method for the synthesis of C_{60} dimers based on the radical reaction of C_{60} with perfluoroalkyl iodides in the presence of $(RSn)_2$ upon irradiation was developed:[22]

$$R_FI \xrightarrow{h\nu,(R_3Sn)_2,\ C_{60}} R_FC_{60}^{\bullet}$$

$$2\ R_FC_{60}^{\bullet} \longrightarrow R_FC_{60}\text{-}C_{60}R_F$$

Preparation of the C_{60} dimers. *A solution of C_{60} (0.05 mmol) and R_FI (0.25 mmol) in 1,2-dichlorobenzene (5 ml) in the presence of $(Bu_3Sn)_2$ (0.25 or 0.50 mmol) was illuminated with a 70 W metal-halide lamp for 5 - 8 h under N_2. The dimers were isolated from the reaction mixture by column chromatography on silica gel followed by gel permeation chromatography (GPC) using JAI model LC-908 liquid chromatography equipped with JAIGEL-1H-40 and 2H-40 columns.*[22]

The synthesis of the dimers $(CH_3)_3C_{60}$–$C_{60}(CH_3)_3$ can be also carried out under oxidation of ${}^tBuC_{60}$ anion with I_2:[23]

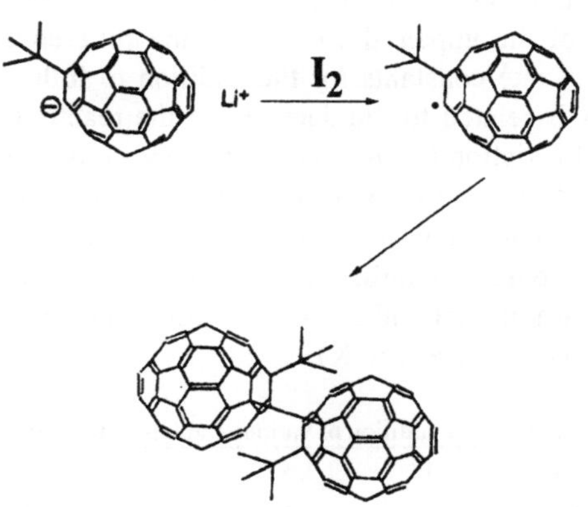

1.1.4 Rate Constants for the Addition of Alkyl Radicals to C_{60}

The rate constants for the addition of various carbon–centered radicals to C_{60} fullerene have been determined with the help of laser flash photolysis and competitive reactions methods.[24 a,b]

Rate constants for the addition of radicals R to C_{60} were found using competitive addition in the presence of different spin traps (α-phenyl-*N*-*tert*-butylnitrone (PBN), 2,3,5,6-tetramethylnitrosobenzene (NB), 2,4,6-tri-*tert*-butylnitrosobenzene (BNB), and 2-methyl-2-nitrosopropane (MNP)).[24 b]

$$C_{60} + {}^\bullet R \xrightarrow{k_1} {}^\bullet C_{60}R, \qquad R = R^1 - R^5$$

R^1 = $^\bullet CM_3$, R^2 = $^\bullet CH_2Me$, R^3 = $^\bullet CH(CH_2)_3Me$, R^4 = $^\bullet CH_2Ph$, R^5 = $^\bullet CH_2CH{=}CH_2$, R^6 = $^\bullet CH(Me)C_2H_5$.

$$S_T + {}^\bullet R \xrightarrow{k_2} {}^\bullet S_TR$$

S_T is a spin trap

The constants obtained have insignificant scatter and are two orders of magnitude higher than the rate constants for the addition to a wide range of monosubstituted unsaturated compounds (Fig. 11).

This higher reactivity of C_{60} compared to monosubstituted alkenes toward carbon-centered radicals is mainly due to the presence of a large number of double bonds in C_{60} and to some energy gain caused by delocalization of the unpaired electron in the fullerenyl radical (Table 3). The decrease in rate constants for the addition to fullerene in the series of R^1–R^6 radicals is related to the decrease in the reactivity of these radicals; similar parallel variation is observed for the rate constants for the addition of these radicals to spin traps or to monosubstituted unsaturated compounds. When the rate constants for addition reactions are known, fullerenes themselves can be used as radical traps functioning at elevated temperatures (450–500 K) (traditional radical traps, nitroso compounds, are thermally unstable and decompose at 350 K).

Table 3. Rate constants for the addition of various carbon-centered radicals to C_{60}.[24 b]

R	$[S_T]_0 x10^3$, mol l^{-1}	$k_1 x10^{-5}$, l mol^{-1} s^{-1}	$k_2 x10^{-5}$, l mol^{-1} s^{-1}
CMe_3	NB; 0.7	174	2000
$MeCH_2$	NB; 0.5	57	400
$MeCH_2$	PBN; 109.5	49	1.3
$Me(CH_2)_4$	PBN; 109.5	49	1.3

R	$[S_T]_0 x 10^3$, mol l^{-1}	$k_1 x 10^{-5}$, l mol^{-1} s^{-1}	$k_2 x 10^{-5}$, l mol^{-1} s^{-1}
$Me(CH_2)_4$	NB; 0.4	41	400
$MeCHC_2H_5$	NB; 0.3	50	400
$MeCHC_2H_5$	MNP; 9.1	46	61
CH_2=$CHCH_2$	NB; 0.2	6.9	400
CH_2=$CHCH_2$	BNB; 6.8	6.0	4.7
$PhCH_2$	NB; 0.1	2.7	57

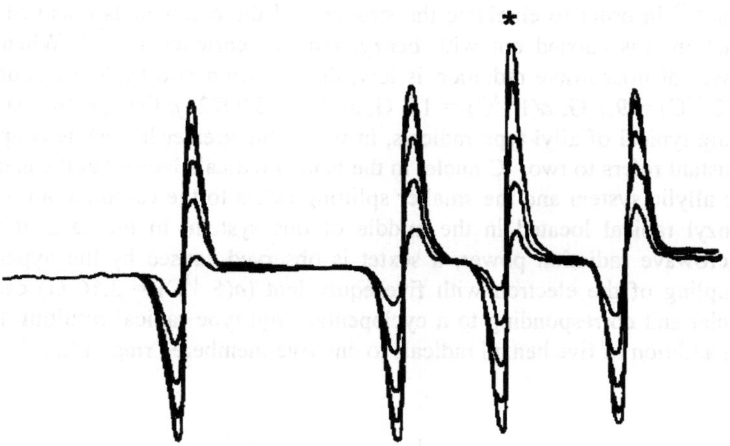

Figure 11. EPR spectrum of the spin-adducts of $Me_3C^•$ with NB (triplet) and C_{60} (singlet *) recorded on exposure to light with $\lambda \sim 366$ nm of a toluene solution containing Me_3CBr, $Re_2(CO)_{10}$, $[NB]_o = 0.2653 \cdot 10^{-3}$ mol l^{-1} and $[C_{60}]_o = 1.0 \cdot 10^{-3}$ mol l^{-1} at 20°C.[24 b]

1.1.5 Multiple Addition of Carbon-Centered Radicals

The products of multiple addition of alkyl, perfluoroalkyl and benzyl radicals to C_{60} fullerenes have been detected.[16,25-27]

1.1.5.1 Multiple Addition of Benzyl Radicals

Prolonged irradiation of a reaction mixture containing C_{60} and a radical source was shown to give rise to compounds with a large number of added species as indicated by line broadening in the EPR spectra and by mass spectra.

Exposure to radiation of a saturated toluene solution of fullerene in the presence of di-*tert*-butyl peroxide brings about an EPR signal whose intensity does not change when irradiation has been terminated, this indicates the formation of a stable adduct.[25]

However, unlike the signals of many carbon-centered radicals, this signal is not saturated upon increase in microwave radiation power even to 200 mW. *g*-Factors measured at 200 mW and at 200 μW are different (2.00221 and 2.00250, respectively) due to the overlap of two signals one of which is saturated upon the increase in the microwave radiation power and the other is not.[25] In order to elucidate the structure of the compounds obtained, the reaction was carried out with benzyl radicals enriched in ^{13}C. When the power of microwave radiation is low, the spectrum is a triplet of doublets ($a(2\ ^{13}C) = 9.7$ G, $a(1\ ^{13}C) = 1.7$ G, and $g = 2.00221$), the spectral pattern being typical of allyl type radicals, in which the greater hyperfine coupling constant refers to two ^{13}C nuclei in the benzyl radicals located at the ends of the allylic system and the smaller splitting refers to the carbon atom of the benzyl radical located in the middle of this system. In the case of high microwave radiation power, a sextet is observed caused by the hyperfine coupling of the electron with five equivalent ($a(5\ ^{13}C) = 3.56$ G) carbon nuclei and corresponding to a cyclopentadienyl type radical resulting from the addition of five benzyl radicals to one five-membered ring in C_{60}:[25]

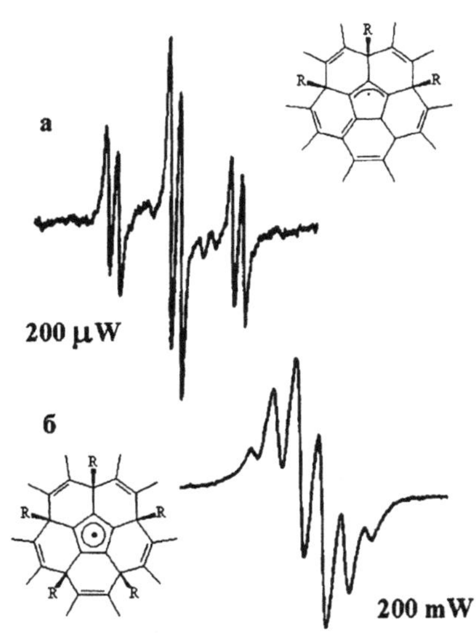

The multiple addition of benzyl radicals to C_{60} can lead to the formation of radical adducts with up to 15 benzyl groups attached to C_{60} (mass spectrometry of the solutions after photolysis with dibenzyl ketone as a source of the benzyl radicals):[25]

$$C_{60} + C_6H_5CH_2(CO)CH_2C_6H_5 \xrightarrow{h\nu} C_{60}(CH_2C_6H_5)_n + CO$$

It was shown, that the use of dimer $[F(C_6H_4)C(CF_3)_2]_2$ as initiator (see Chapter 1) allows the step-by-step controlled addition of the radicals to C_{60}.[26,27] The addition was performed upon dissociation of the dimer (10^{-2} mol l^{-1}) in a saturated solution of C_{60} in toluene at 310 K. The reaction was monitored by EPR and mass spectrometry.

During the first 10 minutes, no signal is observed in the EPR spectrum. However, upon exposure of the reaction mixture to yellow light, a signal appears; when irradiation is terminated, it immediately disappears. This reflects the addition of the first benzyl radical to the fullerene molecule and the subsequent dimerization of two fullerenyl radicals.

The dimer thus formed is diamagnetic; however, the energy of the C–C bond is low and the dimer dissociates to radicals on exposure to yellow light with an energy sufficient for bond rupture (620–680 nm). After that, the addition of second benzyl radical takes place to give a diamagnetic bisadduct:

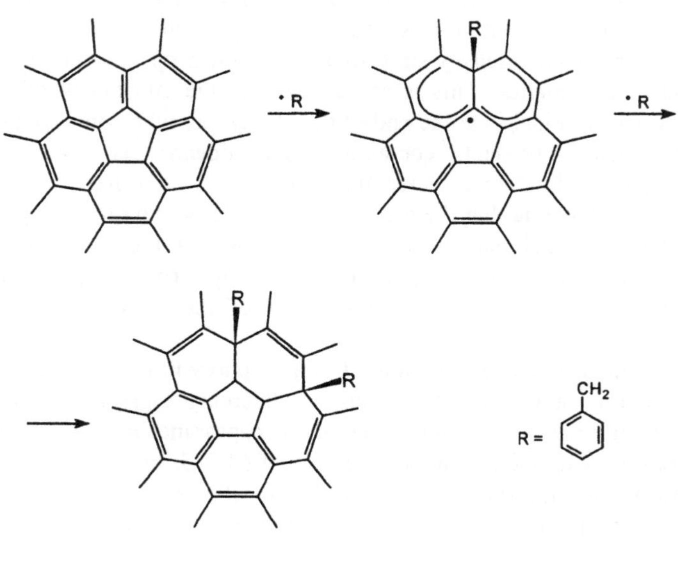

The addition of the third radical can follow two pathways: it either gives an allyl type fullerenyl radical or includes an attack on another position of C_{60}:

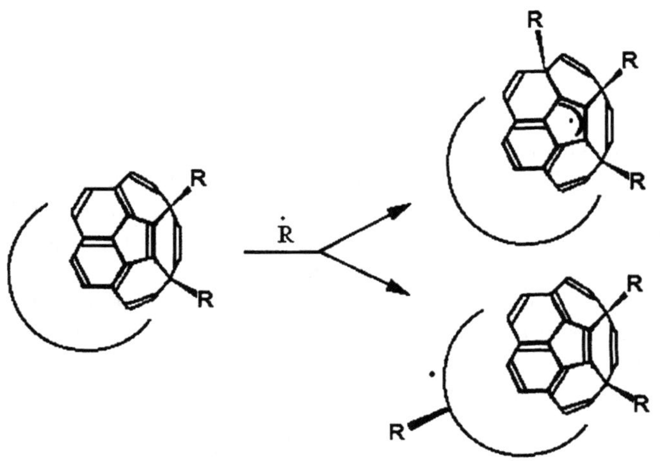

The EPR spectrum displays a signal whose intensity increases for 20 minutes. According to the *g*-factor ($g = 2.0021$), this signal corresponds to the allyl-type fullerenyl radical. Furthermore, it is saturated upon an increase in the microwave radiation power that also supports this conclusion.

Then the signal intensity starts to decrease and drops down to zero over a period of 20 minutes. This is associated with the attachment of a fourth benzyl radical. Finally, at the end of this decay, the tube contains a mixture of diamagnetic products containing four addends. This is confirmed unambiguously by the mass spectrum of the reaction mixture.

It is worth noting that the presence of a minor product resulting from the addition of benzyl radicals to a different site of the fullerene cage is confirmed by means of the phosphoryl EPR spectroscopy procedure (see bellow), which permits the identification of the regiochemistry of the resultant fullerene adducts.[5]

It is known that generation of a phosphorylfullerenyl radical upon irradiation of a mixture of diphosphorylmercury $Hg[P(O)(OPr^{i})_2]_2$ with fullerene in benzene is accompanied by the appearance of a doublet with an *hfc* constant with the phosphorus nucleus of 63.5 G. In the case of addition of phosphoryl radicals to give an allyl-type fullerenyl radical, the constant decreases to 43–49 G. The reaction mixture evaporated after completion of this step was exposed to irradiation in benzene in the presence of Hg[P(O)(OPr^i)$_2$]$_2$; this brought about a doublet with the *hfc* constant at the phosphorus nucleus a_P = 43 G, which corresponds to the allylic radical, and a doublet with a constant of 60 G associated with the addition of a phosphoryl radical to a different position of the fullerene core in the side reaction product and to the major product:[27]

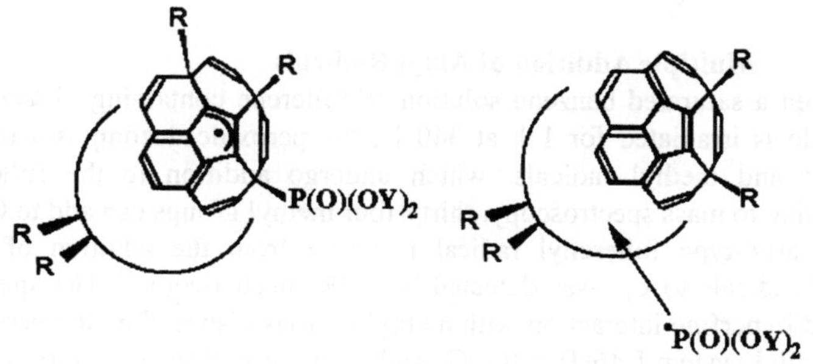

The addition of a fifth benzyl radical is matched by a further increment in the signal intensity; the signal is now formed as a superposition of two signals; judging by their *g*-factors, one signal (g = 2.0021) corresponds to the allylic fullerenyl radical and is saturated upon an increase in the microwave radiation power, and the other signal corresponds to the cyclopentadienyl radical (g = 2.0025):[27]

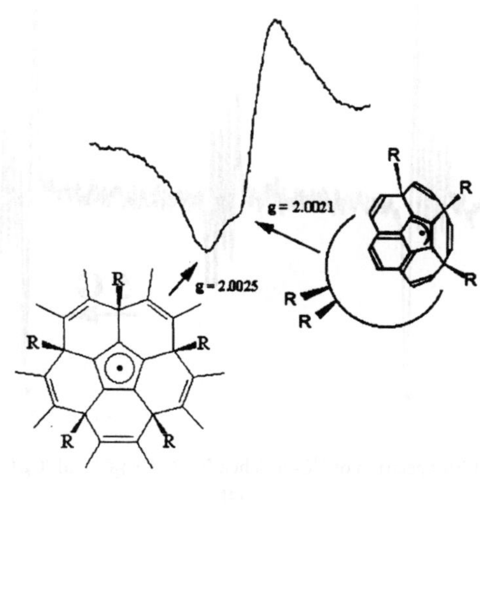

A similar procedure was employed to study the reaction of C_{60} with the 3,5-dimethylphenylmethyl radicals, which are formed, unlike benzyl radicals, even at room temperature. The EPR signal of the allyl-type fullerenyl radical reaches a maximum (after 20 minutes) and then disappears almost completely due to the addition of a fourth 3,5-dimethylphenylmethyl radical. No further increase in the EPR signal intensity, which would point to the addition of a fifth radical, is observed evidently due to steric restrictions caused by the large size of this radical compared to the benzyl radical.[26]

1.1.5.2 Multiple Addition of Alkyl Radicals

When a saturated benzene solution of fullerene containing di-*tert*-butyl peroxide is irradiated for 1 h at 340 K, the peroxide decomposes to give acetone and methyl radicals, which undergo addition to the fullerene. According to mass spectroscopy, thirty-four methyl groups can add to C_{60}.[25]

An allyl-type fullerenyl radical resulting from the addition of three methyl radicals to C_{60} was detected by EPR spectroscopy.[28] The spectrum showed hyperfine interaction with methyl protons disposed on the periphery of the allyl system [a(6H) = 0.2 G] and with the protons of central methyl group [a(3H) = 0.06 G]. The multiple addition of $^{13}CBr_3$ and $^{13}CCl_3$ radicals to C_{60} was studied by EPR spectroscopy. A stable allyl-type fullerenyl radical $^{\bullet}C_{60}(CBr_3)_3$ [a(2C) = 20.1 G, a(1C) = 4.6 G] was detected in the case of tribromomethyl radicals, whereas trichloromethyl radicals produced allylic system whose central position was occupied by Cl atom [a(2C) = 16.4 G, a(1Cl) = 0.61 G (Fig. 12)].[28]

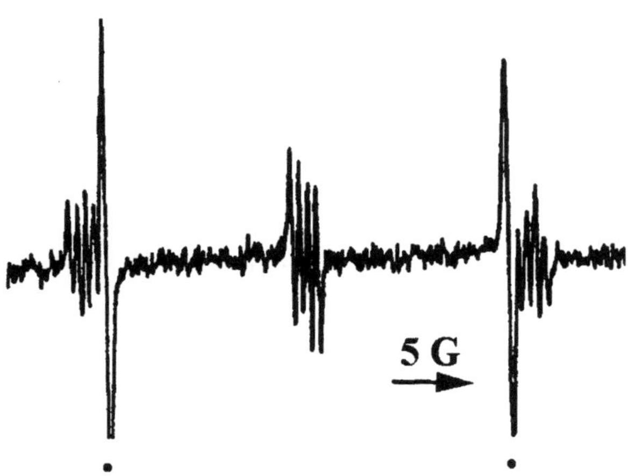

Figure 12. The EPR spectrum of ^{13}C-enriched $^{\bullet}C_{60}(CCl_3)_2Cl$ and $^{\bullet}C_{60}CCl_3$ (•). Adapted from ref 28.

1.1.5.3 Multiple Addition of Fluoroalkyl Radicals

Polyperfluoroalkylated fullerene have been prepared by reaction of fullerenes with a variety of fluoroalkyl radicals, which were generated by thermal or photochemical decomposition of radical precursors, such as fluoroalkyl iodides and fluorodiacyl peroxides.[16]

Heating of C_{60} and excess CF_3CF_2J in 1,2,4-trichlorobenzene at 200 ^{0}C with exclusion of oxygen produced after work-up a dark-brown glassy solid for which electron-capture mass spectrum showed the addition of 6 to 12 CF_3CF_2-groups to C_{60}.

A saturated benzene solution of C_{60} with CF_3J was irradiated with ultraviolet light for 30 min. Electron-capture mass-spectrometry showed addition of up to 13 trifluoromethyl radicals to C_{60}.

The formation of $C_{60}(CF_3)_2H_3^-$ and $C_{60}(CF_3)_3H_4^-$ was detected by negative-ion Fourier transform mass spectrometry. Such products could form on account of hydrogen abstraction by fullerenyl radicals—$CF_3C_{60}^•$, $(CF_3)_2HC_{60}^•$, $(CF_3)_3H_2C_{60}^•$, etc.—from the solvent or from the adducts of $CF_3^•$ with benzene. To exclude this possibility, trifluoromethylation was carried out by heating of a heterogeneous mixture of C_{60} and excess of CF_3J in hexafluorbenzene at 200 °C.[16]

Addition of trifluoromethyl groups to C_{60} increases the solubility of the resultant adducts. According to mass spectra, the addition of 14 trifluoromethyl radicals was achieved after heating during 24 h. Similar results were obtained with perfluoroethyl, perfluoropropyl, and perfluorohexyl iodides.

In the spectrum of the stable fullerenyl radical of the allyl type, the product of the addition of three trifluoromethyl groups was detected, which shows *hfi* only with fluorines of two last CF_3-groups (a(6F) = 0.36 G).[28]

Other sources of fluoromethyl radicals can be also used. Reaction of C_{60} with stable perfluoroalkyl radical[29,30] $[(CF_3)_2CF]_2{}^•CCF_2CF_3$ (E_{act} of decay ~ 30 kcal mol^{-1}) in 1,2,4-trichlorobenzene at 400 K affords a mixture of trifluoromethylated products containing up to 18 trifluorometyl addends (Fig. 13):[31]

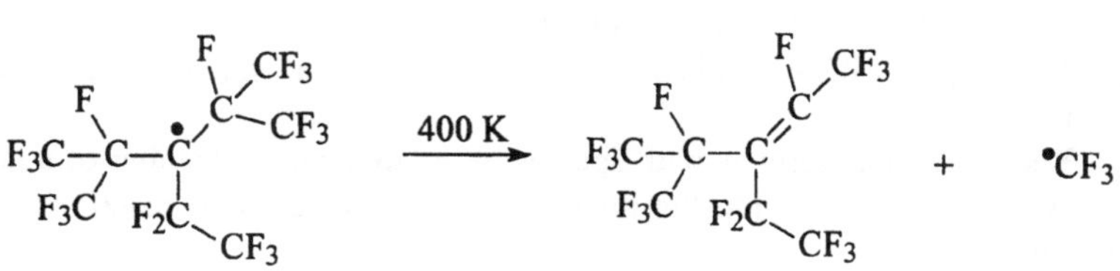

$$C_{60} + n\ {}^{\bullet}CF_3 \rightarrow C_{60}[CF_3]_n$$

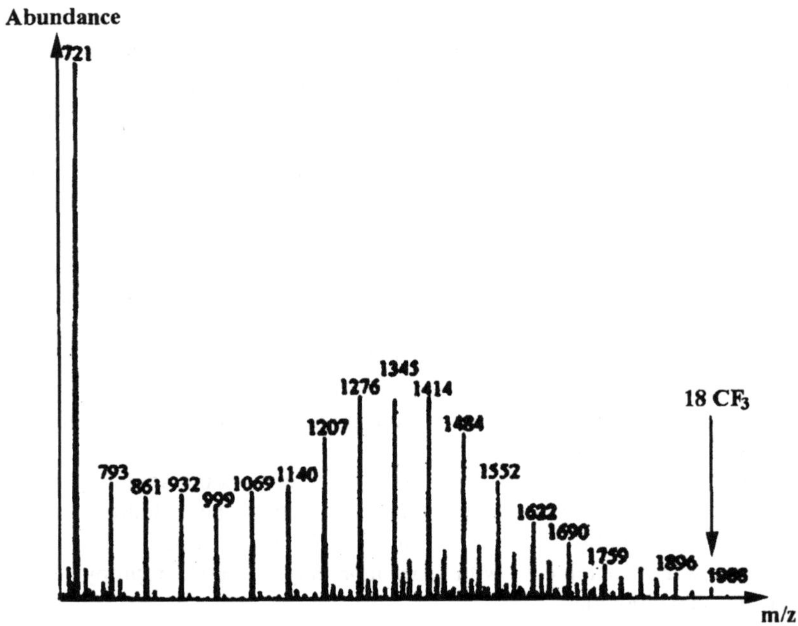

Figure 13. High mass region of EI mass spectrum of the products from the reaction of C_{60} with $[(CF_3)_2CF]_2{}^{\bullet}CCF_2CF_3$ at 200 °C.

Simultaneous decrease in the intensity of $[(CF_3)_2CF]_2{}^{\bullet}CCF_2CF_3$ and increase in the intensities of the radical adducts ${}^{\bullet}C_{60}(CF_3)_n$ (n is an odd number) is observed when this reaction is carried out in the resonator of an EPR spectrometer (Fig. 14).

The trifluoromethylation of C_{60} and C_{70} with the use of $[(CF_3)_2CF]_2{}^{\bullet}CCF_2CF_3$ radical was studied in a mass spectrometer chamber.[31a]

Reactions of fullerenes with species in the ionization chamber of a mass spectrometer may provide the possibility to carry out such reactions on a preparative scale.[31b]

The reaction occurs at least partly on the walls of the ionization chamber by a radical mechanism with participation of CF_3 radicals resulting from the radical reactants *via* thermal decomposition or under electron impact:[31a]

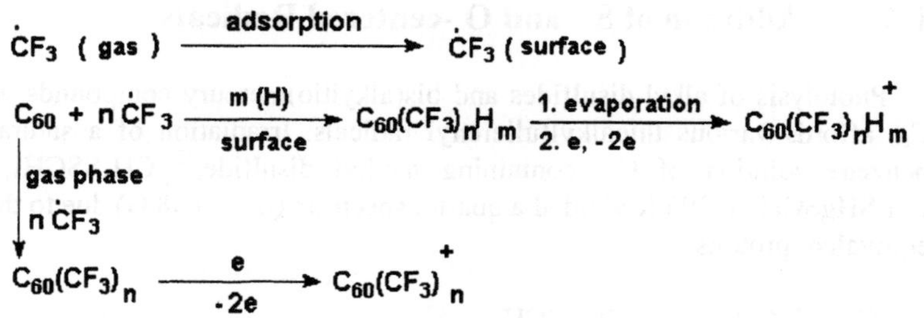

Reaction of C_{60} with a solution of perfluoropropionyl peroxide, C_2F_5-C(O)O-O(O)C-C_2F_5, at 25 °C produced clear dark-brown solutions. A typical mass spectrum of the product showed from 9 to 16 perfluoroethyl radicals attached to C_{60}.[16]

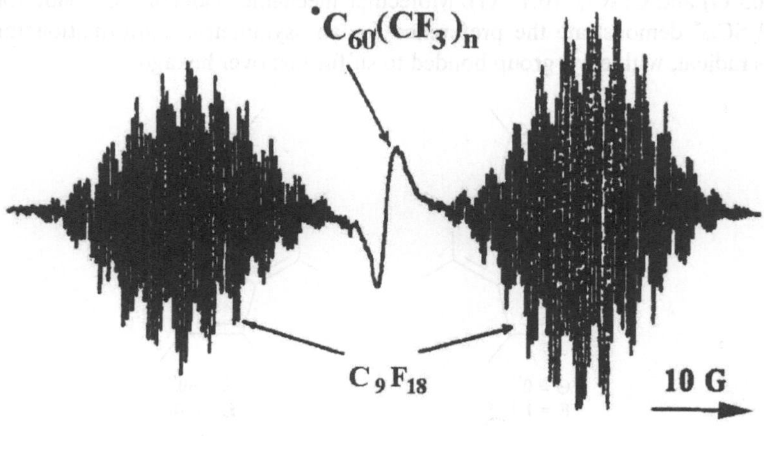

Figure 14. EPR spectra of $[(CF_3)_2CF]_2{}^{\bullet}CCF_2CF_3$ and the polyaddition spin-adducts of $^{\bullet}CF_3$ radicals to C_{60}.[27]

Polyperfluoroalkylated fullerene appears interesting physical properties. The materials obtained are stable up to 270 °C. Under high vacuum, the perfluorohexylated nanospheres can be sublimed quantitatively to deposit a thin film on a glass slide.[16]

Radical reactions of fullerenes with different element-centered radicals were also studied.

1.2 Addition of S - and O -centered Radicals

Photolysis of alkyl disulfides and bis(alkyltio)mercury compounds with C_{60} affords various thioalkylfullerenyl radicals. Irradiation of a saturated benzene solution of C_{60} containing methyl disulfide,[32] CH_3SSCH_3, or $CH_3SHgSCH_3$ at 295 K yielded a quartet spectrum ($a_H = 0.38$ G) due to three equivalent protons:

$$C_{60} + {}^{\bullet}SCH_3 \longrightarrow {}^{\bullet}C_{60}SCH_3$$

The g-factor (2.0024) of the quartet spectrum is higher than the values observed for monoalkyl radical adducts (g = 2.0022). A positive g-factor shift is expected for radicals with spin density on sulfur because of its relatively large spin-orbital coupling. The *hfi* of δ-protons in $CH_3SC_{60}^{\bullet}$ is substantially larger than those of γ-protons and even γ-fluorines in $CH_3C_{60}^{\bullet}$ (0.03 G) and $CF_3C_{60}^{\bullet}$ (0.13 G). Molecular mechanics calculations made for $CH_3SC_{60}^{\bullet}$ demonstrate the preference for an asymmetric conformation for this radical, with ethyl group bonded to sulfur just over hexagon:[32]

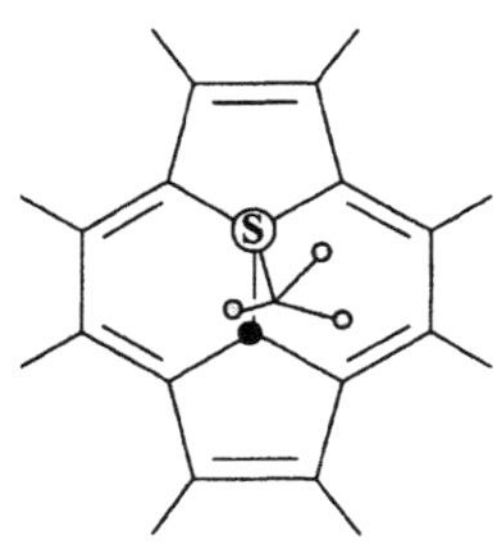

Θ = 0°
E = 1.1

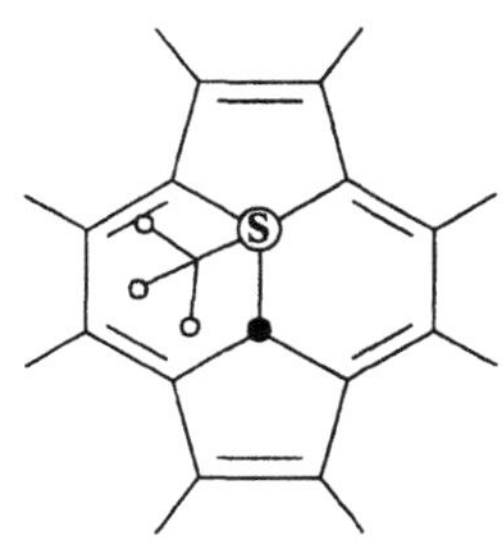

Θ = 60°
E = 0.0

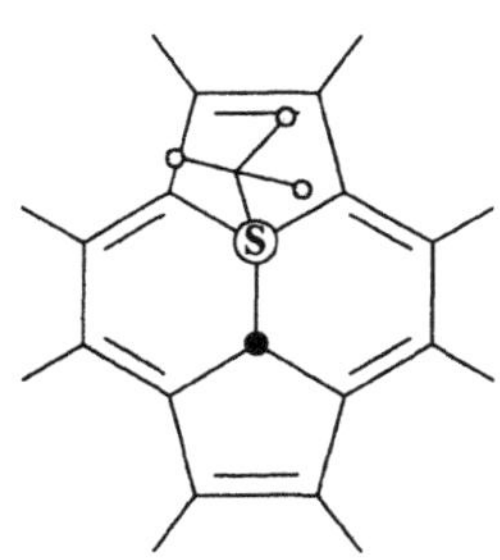

Θ = 140°
E = 1.2

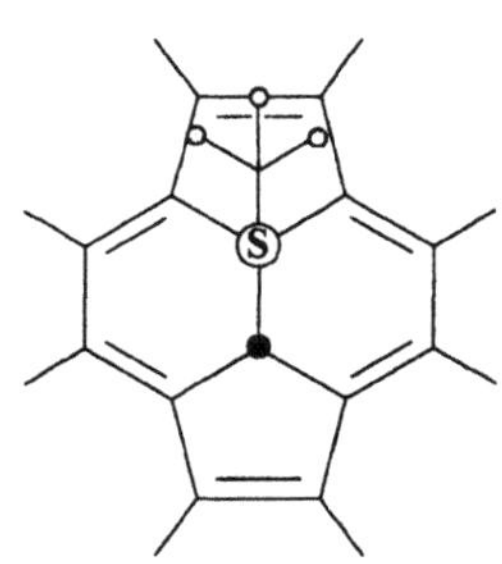

Θ = 180°
E = 1.1

In contrast to the spectra of alkylfullerenyl radicals, which become more intense at higher temperatures owing to the dissociation of $RC_{60}–C_{60}R$ dimers, the spectrum of $CH_3SC_{60}^{\bullet}$ rapidly decreases in intensity above room temperature. This difference in behavior is associated with the reversibility of the alkylthio radical addition due to a weaker fullerene–sulfur bond rather than with the lack of dissociation of the corresponding $RSC_{60}–C_{60}SR$ dimers, which should result from these photochemical reactions by analogy.[32]

Irradiation of phenyl disulfide gives a sulfur-centered fullerenyl radical that is accompanied by a change in colour of the reaction mixture from purple to brown. Within 30 min after irradiation ceased, the purple color is restored, indicating that the radical addition is reversible; this has not been observed for alkoxy fullerenyl derivatives. Evidently, this is due to the lower strength of the C–S bond and its easier dissociation.[32]

The spectral parameters of other thio radical adducts are listed in the Table 4. The ethylthio adduct of C_{60} should be noted, where all protons are almost equivalent.

Alkoxy radicals can also add to C_{60}.[32,33] The spectrum of the *tert*-butoxy adduct obtained by photolysis of a saturated benzene solution of C_{60} containing $^tBuOOBu^t$ shows the splitting for the nine equivalent protons (0.35 G) which is larger than that of the nine equivalent protons of the *tert*-butyl adduct (0.17 G). Moreover, an adduct of methyl radical also formed due to decay of *tert*-butoxy radical.

An analogous experiment with the use of CF_3OOCF_3 afforded a 1 : 3 : 3 : 1 quartet with unusually large coupling (3.14 G) with the three equivalent fluorines, quite distinct from that of the CF_3 radical adduct (0.13 G), which can be assigned to $CF_3OC_{60}^{\bullet}$.[33]

UV irradiation of dialkoxy disulfides ROSSOR (R = Me, Et) in the presence of C_{60} resulted in the formation of methoxy-, ethoxy-, and *iso*-propoxyfullerenyl adducts. An interesting feature of the EPR spectra of these compounds is the absence of splitting at the hydrogen atoms linked to the carbon that occupies the α–position with respect to oxygen. The spectrum of $MeOC_{60}^{\bullet}$ displays a single line whereas the hydrogens of the methyl groups in the β-position to oxygen yield a 4-line, a 7-line, and a ten-line multiplet in the case of $EtO\text{-}C_{60}^{\bullet}$, $i\text{-}PrOC_{60}^{\bullet}$, and $^tBuO\text{-}C_{60}^{\bullet}$, respectively. Furthermore, there was no *hfi* with two protons in the β-position to the oxygen in the $PhCH_2OC_{60}^{\bullet}$.

One can suppose the difference in the results obtained for $EtSC_{60}^{\bullet}$ and $EtOC_{60}^{\bullet}$ to be associated with the different geometry because the C–S bond is longer than the C–O bond. Molecular mechanics calculations show that the CH_3 and CH_2 groups occupy a position where both groups can similarly overlap with the p_z orbital bearing the unpaired electron in the sulfur-containing radical, whereas only the CH_3 group can experience such an interaction in the case of $EtOC_{60}^{\bullet}$:

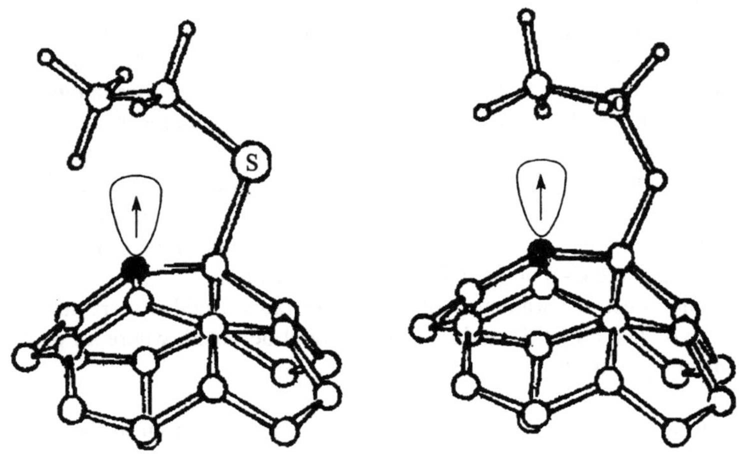

$^{\bullet}C_{60}SEt$ $^{\bullet}C_{60}OEt$

Table 4. Hfc constants for O- and S-fullerenyl radicals.

$R^{\bullet}$	g-factor	a_{Me}, G	a_{CH2}, G	a_{CH}, G	Ref.
SH_3		3 H = 0.38			33
$SCHMe_2$		6H = 0.22		1 H = 0.22	32
SCH_2CH_2Me			2 H = 0.35 2 H = 0.18		33
SCH_2CHMe_2			2 H = 0.40	1 H = 0.21	32
$SCMe_3$		9 H = 0.25			32
SEt		3 H = 0.29	2 H = 0.31		32
SMe	2.0024	0.38			32
$OCMe_3$		9 H = 0.35			32
$OCMe_2Ph$		6 H = 0.22 3 H = 0.17 (Ph)			33
OCF_3		3 F = 3.14			33
OMe	2.0023				32
OEt		3 H = 0.37			32
OPr^i		6 H = 0.27			32
OPr^iCH_2		6 H = 0.15		1H = 0.47	32

The stability and reactivity of polyadducts of *tert*-butoxy radicals to C_{60} were studied.[34] $(Bu^tO)_nC_{60}^{\bullet}$ radicals have been produced by addition of photochemically generated $Bu^tO^•$ to C_{60}. Radical adducts (at $n > 2$) are very persistent under N_2 at room temperature and can be observed for several days. Reactions between the radical adducts and O_2 are slow. However, nitric oxide reacts more rapidly with stable fullerenyl radicals. Authors decided that difullerene aminoxyl radicals $[(Bu^tO)_nC_{60}]_2N\text{-}O^•$ can form in this reaction, which have been observed by EPR spectroscopy at concentrations of *ca.* 10^{-7} M:

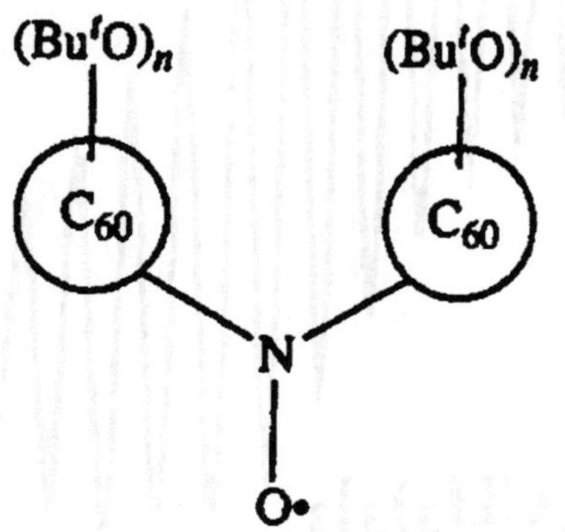

As it follows from their nature, $(Bu^tO)_nC_{60}^•$ radicals are not reactive toward certain typical radical-trapping agents, *e.g.*, 5,5-dimethyl-1-pyrroline N-oxide, 2-methyl-2-nitrosopropane, etc.[34]

1.3 Addition of Silyl Radicals

Radical adducts of silyl radical with C_{60} were also studied (Table 5):[9]

$$R_3Si^• + C_{60} \longrightarrow R_3SiC_{60}^•$$

The corresponding silanes were used as a source for silyl radicals. The spectrum of $Me_3Si–C_{60}^•$, like that for $Me_3C–C_{60}^•$, has hyperfine structure at 300 K due to nine equivalent protons. The $Si–C_{60}$ bond is slightly longer than the $C–C_{60}$ bond in $Me_3C–C_{60}$. As a result, the barrier of rotation about $Si–C_{60}$ is lower. This could be the reason why the hindered rotation in $Me_3Si–C_{60}^•$ was not observed in the EPR spectrum (Fig. 15). In the EPR spectrum of $(Me_3C)_3Si–C_{60}^•$, the interaction of the unpaired electron with 27 equivalent protons can be seen clearly. The analysis of the EPR spectra of

$Et_3Si–C_{60}^{\bullet}$ and $^iPr_3Si–C_{60}^{\bullet}$ showed that free rotation around $Si–C_{60}$ bond takes place, although the Si–R bonds are non-rotating on the EPR time scale.

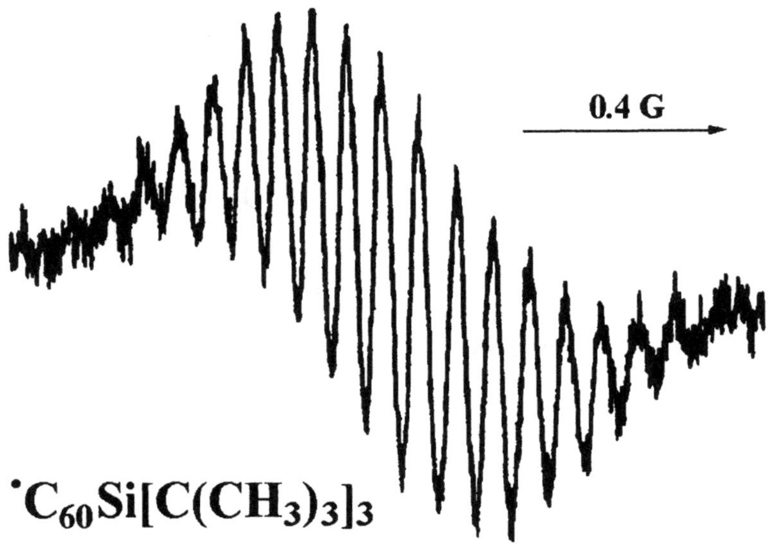

Figure 15. The EPR spectrum of $C_{60}SiBu^t_3$. Adapted from ref 9.

Table 5. Hfc constants for Si-fullerenyl radicals.

R	a_H, G	T, K	Ref.
$SiMe_3$	9 H_δ = 0.11	320	9
$SiEt_3$	3 H_δ = 0.30	320	9
	3 H_δ = 0.025		
	9 H_ε = 0.15		
$Si(Me_2CH)_3$	3 H_δ = 0.28	295	9
	9 H_ε = 0.14		
$Si(CMe_3)_3$	27 H_ε = 0.075	310	9

The adducts of an unusual structure were obtained by reacting aryl-substituted silyl radicals with C_{60}.[35] The photolisys of unsymmetrical disilanes in the presence of C_{60} gave rise to the adducts of two main types (1,16- and 1,2-adducts):

The authors suppose that the fullerenyl radical formed at the first step of the addition of the phenyl substituted silyl radical undergoes intramolecular attack to the aromatic ring to produce the cyclohexadienyl radical. This radical is in equilibrium with the fullerenyl radical. The interaction of the next silyl radical with the cyclohexadienyl radical leads to the formation of the adduct of the 1,2-type; its addition to fullerene gives the 1,16-adduct as shown bellow:[35]

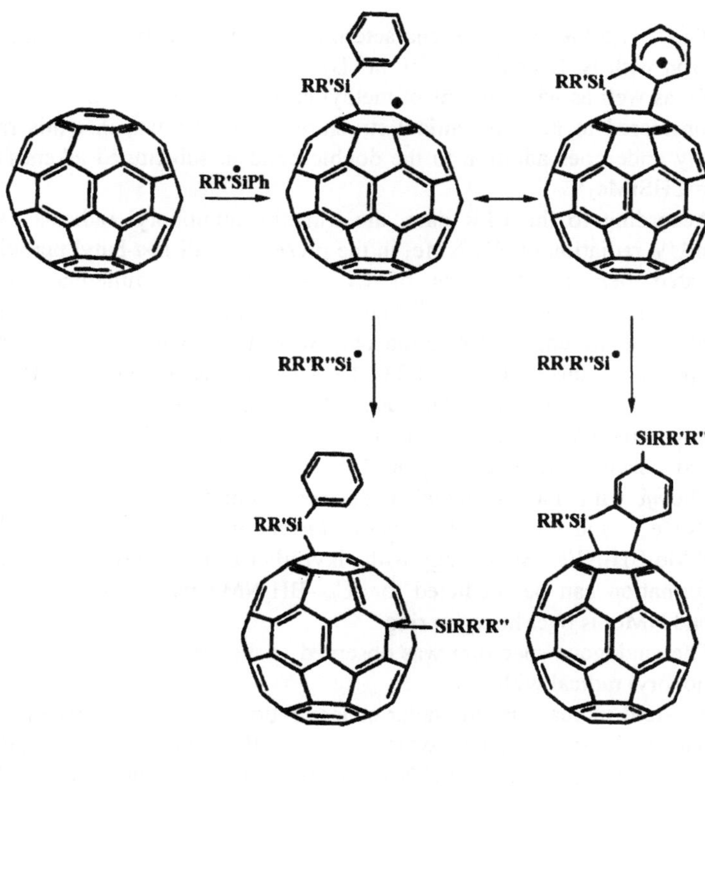

1.4 Addition of Boron-Centered Radicals

The nuclei of boron isotopes (^{10}B, ^{11}B) possess magnetic moments and the hyperfine coupling of the unpaired electron with these nuclei is clearly manifested in the EPR spectra.

The addition of boroncentered radicals derived from BH_3NMe_3 and carboranyl radicals to C_{60} was studied.[36,37]

1.4.1 Addition of Boryl Radicals

For the first time, the trimethylaminoboryl radicals were obtained as follows:[38]

$$Bu^tO{-}OBu^t \xrightarrow{h\nu} 2\ {}^{\bullet}OBu^t$$

$$BH_3NMe_3 + {}^{\bullet}OBu^t \longrightarrow {}^{\bullet}BH_2NMe_3 + HOBu^t$$

The ${}^{\bullet}BH_2NMe_3$ radical is characterized by the interaction of the unpaired electron with B, H and N nuclei: $a(^{11}B) = 51.3$ G, $a(2H) = 9.6$ G, $a(^{14}N) = 1.4$ G, as well as with protons of methyl groups $a(6H) = 1.4$ G. This radical is considered to have pyramidal structure. It was shown, that this radical readily undergoes addition to the double bond in substituted alkenes (e.g., $CH_2{=}CHSiMe_3$).

According to the EPR data, the trimethylaminoboryl radicals formed upon UV radiation of BH_3NMe_3 in the presence of di-*tert*-butyl peroxide in saturated benzene solutions of C_{60} combine with fullerene to give monoadducts ${}^{\bullet}C_{60}{-}BH_2NMe_3$.[36] The EPR spectrum of this radical exhibits coupling of the unpaired electron only with the magnetic nuclei of boron isotopes ($g = 2.0018$, $a(^{10}B) = 4.2$ G, $I = 3$, seven lines with the intensities 1 : 1 : 1 : 1 : 1 : 1 : 1 (natural abundance 18.8%); $a(^{11}B) = 12.4$ G, $I = 3/2$, four lines with the intensities 1 : 1 : 1 : 1 (natural abundance 81.2 %)), which can be easily seen in the spectrum (Fig. 16).

The *hfc* with H and N nuclei is not observed in the EPR spectrum of ${}^{\bullet}C_{60}$-BH_2NMe_3, and it is two times lower with ^{11}B than in ${}^{\bullet}CH_2$-$CHSiMe_3BH_2NEt_3$. By analogy with alkylfullerenyl radicals, unsymmetrical conformation can be predicted for ${}^{\bullet}C_{60}{-}BH_2NMe$ where more nucleofilic addend NMe_3 is over hexagon ring.

The analogous spectrum was observed in the reaction of the deuterated aminoboryl radical with C_{60}.

A 5-fold increase in the molar concentration of BH_3NMe_3 results in the formation of a stable radical, which is apparently due to multiple addition of the boryl radicals to fullerene. No concerted addition giving rise to allyl or cyclopentadienyl radical is observed. Boryl radicals add to C_{60} positions which are more distant from the carbon atom bearing the maximum spin density. Therefore, *hfc* of the unpaired electron with several remote nuclei results in line broadening in the EPR spectrum (Fig. 16b). Yet another possible reason for this broadening is a decrease in the speed of rotation of the radical owing to the increase in its weight accompanied by an increase in the contribution of *g*-factor anisotropy and *hfc* to the line width.

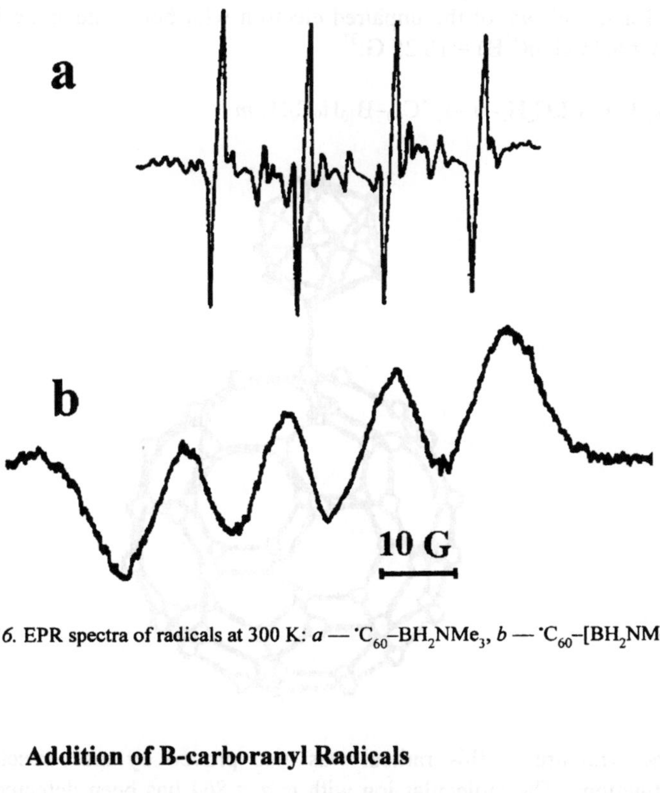

Figure 16. EPR spectra of radicals at 300 K: *a* — $^{\bullet}C_{60}$–BH_2NMe_3, *b* — $^{\bullet}C_{60}$–$[BH_2NMe_3]_n$.[36]

1.4.2 Addition of B-carboranyl Radicals

The radical adducts of polyhedral boron-centered *o*-, *m*-, and *p*-carboranyl-12 radicals were first detected when 4-methyl-2,4,6-tri-*tert*-butylcyclohexa-2,5-dien-1-one was employed as a spin trap.[39] It has been shown that photolysis of di(*m*-carborane-9-yl)mercury solution results in its decomposition to form B-carboranyl radicals, and this route is more favorable than hydrogen abstraction from the boron atom with *tert*-butoxyl radical.

The EPR spectra of the radical adducts of carboranyl radicals with 4-methyl-2,4,6-tri-*tert*-butylcyclohexa-2,5-dien-1-one are characterized by *hfc* constants depending on the structure of the carborane molecule (^{11}B = 25.2 G (*o*-carborane), 25.7 G (*m*-carborane), 26.5 G (*p*-carborane)).

The addition of *m*-carboranyl radicals to the branched fluoroalkenes leads to the formation of a stable radicals with *hfc* $a(^{11}B)$ = 15 G.

UV irradiation of $Hg[B_{10}H_9C_2H_2\text{-}m]$ in a saturated toluene solution of C_{60} gives rise to the EPR spectrum of B–carboranylfullerenyl radical with g = 2.0023 and with *hfc* of the unpaired electron with boron magnetic isotopes $a(^{10}B)$ = 6.25 G, $a(^{11}B)$ = 18.25 G:[37]

$$C_{60} + {}^{\bullet}B_{10}H_9C_2H_2\text{-}m \rightarrow {}^{\bullet}C_{60}\text{–}B_{10}H_9C_2H_2\text{-}m$$

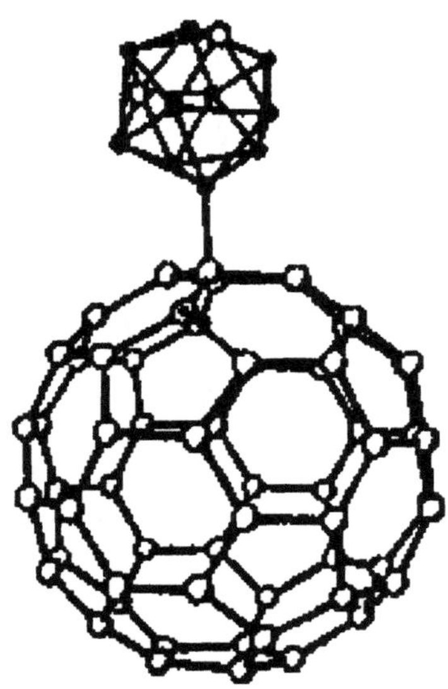

The structure of this radical was also proved by mass-spectrometry investigations. The molecular ion with m/z = 864 has been detected due to the addition of one B-carboranyl radical to the fullerene. In spite of relatively large volume of carborane substituent, the carboranylfullerenyl radicals dimerize after irradiation ceased that was confirmed by the following experimental facts:

1. The proportionless dependence of signal intensity on irradiation intensity;
2. The steady-state concentration of ${}^{\bullet}C_{60}\text{–}B_{10}H_9C_2H_2\text{-}m$ radical increases with temperature above 300 K;

3. The irradiation of sample at 620–680 nm gives rise the dissociation of dimer RC_{60}–$C_{60}R$ (R = $B_{10}H_9C_2H_2$-*m*). This effect is observed after 10–12 min UV irradiation of Hg[$B_{10}H_9C_2H_2$-*m*]$_2$ in a saturated toluene solution of C_{60}, which leads to accumulation of dimer RC_{60}–$C_{60}R$. No radicals were detected if the sample was initially irradiated with visible light ($\lambda > 620$ nm) since such light does not cause the decomposition of the mercury compound.

The enthalpy of equilibrium (ΔH) was estimated from dependence $\ln C^2$ *vs*. 1/T, where *C* is radical concentration:

$$2\ {}^{\bullet}C_{60}B_{10}H_9C_2H_2\text{-}m \Leftrightarrow m\text{-}H_2C_2H_9B_{10}C_{60}\text{–}C_{60}B_{10}H_9C_2H_2$$

The obtained value $\Delta H \sim 10$ kcal/mol is lower than those for the majority of dimers of alkylfullerenyl radicals that seems to be due to steric hindrances to dimerization imposed by the bulky carborane fragment. The dimerization rate constant for ${}^{\bullet}C_{60}$–$B_{10}H_9C_2H_2$-*m* radical measured by kinetic EPR spectroscopy at 0 °C was found to be $\sim 10^6$ l mol^{-1} s^{-1}.

Prolonged photolysis (30 min) of a saturated toluene solution containing C_{60} and excess of Hg[$B_{10}H_9C_2H_2$-*m*]$_2$ results in multiple addition of carboranyl radicals to fullerene. In this case a stable spin-adduct of multiple addition is detected in the EPR spectrum for which the spectral lines are ~5 times broader. This is caused by superposition of the EPR signals from the spin-adducts of polyaddition of B-carboranyl radicals to C_{60} with various numbers of radicals. This has been also proved by mass-spectrometry (Fig. 17).[40]

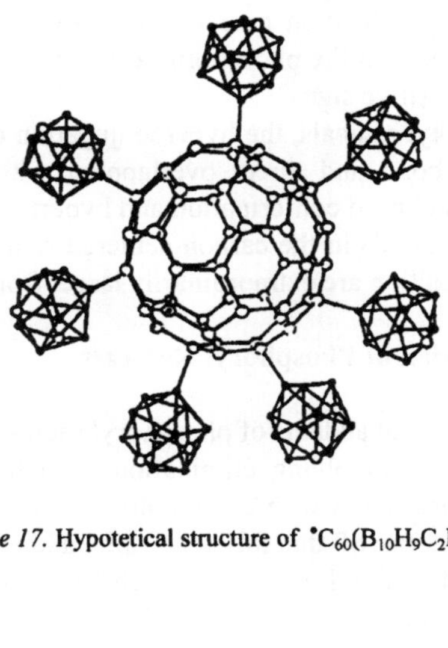

Figure 17. Hypotetical structure of ${}^{\bullet}C_{60}(B_{10}H_9C_2H_2)_7$.[40]

1.5 Addition of Phosphoryl Radicals

Depending on the coordination number of phosphorus in the compound, phosphinyl ($^{\bullet}PR_2$),[41] phosphoryl $^{\bullet}P(O)R_2$,[42] and phosphoranyl radicals ($^{\bullet}PR_4$)[43] are distinguished. Due to the exceptionally high constants of hyperfine coupling of the unpaired electron with the phosphorus nucleus, the two last-mentioned types of phosphorus-centered radicals are the most promising "paramagnetic reporters" (for $^{\bullet}P(O)R_2$, a_P = 400—700 G and for ($^{\bullet}PR_4$), a_P = 500—1300 G). Indeed, it is for phosphoranyl radicals that diastereomeric anisochronism was first detected by EPR spectroscopy when the *hfc* constants for the two diastereomers differed by only 1%.[44] However, most phosphoranyl radicals are thermally unstable and decompose. Apart from phosphoranyls derived from cyclic phosphites, only phosphoryl radicals exhibit a tendency of adding to double bonds.

Owing to the large a_P values for the spin adducts of phosphoryl radicals (40–100 G) in which the phosphorus atom occupies a β-position with respect to the radical center, EPR spectra can provide information on the intramolecular interactions in the radical, the influence of the closest environment of phosphorus on the rate of conformation transitions, and, finally, on the regiospecificity of the addition of phosphoryl radicals. For example, EPR has been employed to study the addition of phosphoryl radicals to C=C, C=S, and C=O bonds and their reactions with internal perfluorinated olefins, which are direct analogs of fullerenes, as noted above, highly electron-acceptor molecules.[45-47]

Phosphoryl radical has a pyramidal structure with the unpaired electron on the orbital with hybridization near to sp^3. Large contribution of the s orbital in the spin density of the phosphorus atom allows one to suggest that the spin density has positive sign.

In β-phosphorylethyl radicals, the hyperconjugation of unpaired electron with C_β–$P(O)(OR)_2$ bond and direct overlapping with 3s or 3p orbitals increases stability of eclipsed conformation and hyperfine interaction, $a(^{31}P)$. Indeed, the values of $a(^{31}P)$ in the carbon-centered radical with phosphoryl substituent at the β position are extraordinarily large: from 45 to 140 G.[42,47,48]

1.5.1 Monoaddition of Phosphoryl Radicals

The synthesis of radical adducts of phosphoryl radicals with fullerenes by the standard procedure, involving elimination of a hydrogen atom from dialkyl phosphite on treatment with a *tert*-butoxyl radical, does not produce very intense EPR spectra of the fullerene spin adducts with phosphoryl radicals[8,16] because fullerenyl radicals detach hydrogen from the initial hydrophosphoryl compound.[49]

Therefore, to prepare phosphorylfullerenyl radicals, one can use photolysis of diphosphorylmercury compounds; previously, it had been shown that they are efficient sources of phosphoryl radicals.[50]

$$Hg[P(O)(OR)_2]_2 \xrightarrow{h\nu} 2\ {}^{\bullet}P(O)(OR)_2 + Hg$$

$$C_{60} + {}^{\bullet}P(O)(OR)_2 \longrightarrow {}^{\bullet}C_{60}P(O)(OR)_2$$

R = Me, Et, *i*-Pr

Synthesis of bis-(dimethoxyphosphinyl)-mercury. *To a vigorously stirred solution of dimethyl phosphonate (50 g, 0.45 mole) in 75 ml of dry benzene, was added 49.0 g (0.23 mole) of mercuric oxide at room temperature, followed by additional 50 ml of benzene after the oxide addition was complete. The pressure was reduced to 200 mm and the reaction mixture was refluxed at 40-45^0. The water produced by the reaction was removed as the benzene azeotrope and collected in a Dean-Stark trap. The product precipitated from the benzene solution on cooling. After one recrystalization from benzene, 78.9 g (83% yield) of bis-(dimethoxyphosphinyl)-mercury was obtained.*[50]

When $Hg[P(O)(OR)X]_2$ is photolyzed in a saturated solution of C_{60} in toluene or benzene, EPR spectra of spin adducts are recorded; the *hfc* constants of the unpaired electron with the ^{31}P nuclei are 63.5 G (${}^{\bullet}C_{60}P(O)(OPr^i)_2$) and 63.7 G (${}^{\bullet}C_{60}P(O)(OEt)Ph$) (Fig. 18).[51]

Analysis of the *hfc* constants of the unpaired electron with the ^{13}C magnetic isotopes in alkylfullerenyl radicals and comparison of these data with the corresponding values for radicals ${}^{\bullet}C_{60}P(O)(OPr^i)_2$ ($a^{13}C$ = (1C) = 18.5 G; (4C) = 9.0 G; (3C) = 7.2 G; (2C) = 6.0 G; (6C) = 3.6 G; (8C) = 2.0 G) indicates that, in the phosphorylfullerenyl radicals, the unpaired electron is mainly delocalized over two six-membered rings adjacent to the ${}^{\bullet}C$–$CP(O)(OR)_2$ bond, as in alkylfullerenyl radicals.[10]

The EPR spectra of phosphorylfullerenyl radicals were studied at low temperatures (in the 273—193 K range).[52] It was found that as the temperature decreases and the rate of rotation of the phosphoryl group around the C–P bond decreases, the lines are first broadened and then two doublets appear due to the two "limiting" conformations differing in both the *hfc* constants of the unpaired electron with the ^{31}P nucleus and *g*-factors (Fig. 19).[52]

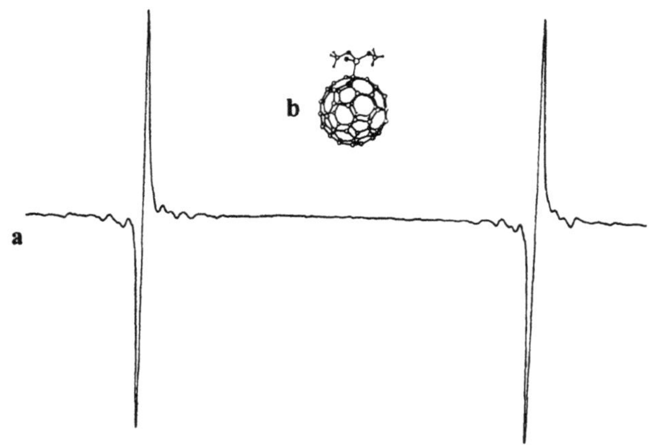

Figure 18. EPR spectrum of $^{\bullet}C_{60}P(O)(OPr^{i})_2$ at 290 K (a) and its structure (b).[51]

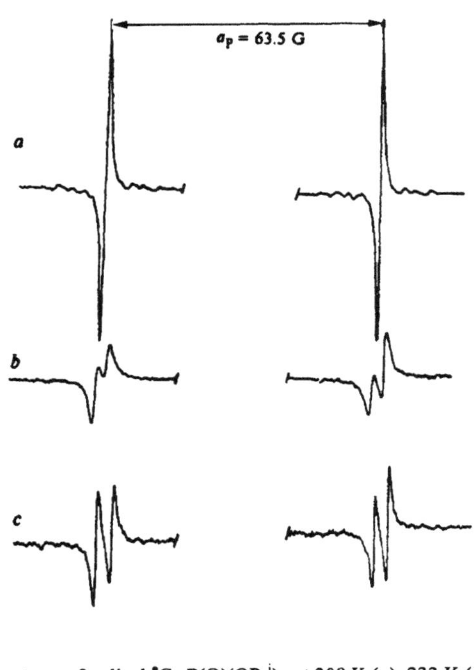

Figure 19. EPR spectrum of radical $^{\bullet}C_{60}P(O)(OPr^{i})_2$ at 300 K (*a*), 233 K (*b*), and 200 K (*c*).[52]

The spectral and thermodynamic data obtained make it possible to estimate the rate constants (k_1 and k_{-1}) for the transition between the two conformations in the temperature range (270—240 K).

The constant (k_{-1}) as a function of temperature is as follows (s^{-1}): k_{-1} (270 K) = $5 \cdot 10^7$; k_{-1} (260 K) = $3.8 \cdot 10^7$; k_{-1} (250 K) = $2.7 \cdot 10^7$; and k_{-1} (240 K) = $1.1 \cdot 10^7$. The activation energy of this process was found to be 4.8 kcal mol^{-1}. The value of k_1 at 270 K is $4 \cdot 10^6$ s^{-1}:[52]

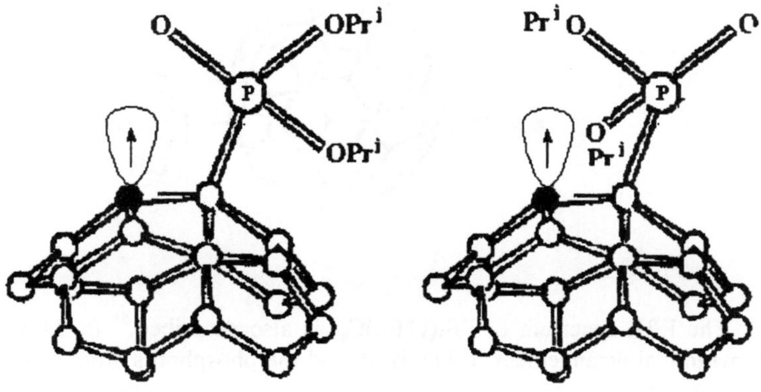

It was shown that the magnetic-resonance parameters of the spectra can be affected by the addition to the reaction mixture of fluorinated alcohols capable of interaction with the phosphoryl group.[53] The alcohols used were CF_3CH_2OH, $(CF_3)_3COH$, and $(CF_3)_2CHOH$.

In the EPR spectrum of radical $^{\bullet}C_{60}P(O)(OPr^i)_2$ as a complex with CF_3CH_2OH, the *hfc* constant with the phosphorus nucleus (a_P = 66.8 G, g = 2.0022) is 3.3 G greater than that for the non-complexed spin adduct of the phosphoryl radical with C_{60}, a_P = 63.5 G, g = 2.0023.

The *hfc* constant with the phosphorus nucleus in the phosphorylfullerenyl radical increases to even a greater extent upon the addition of the alcohols $(CF_3)_3COH$ and $(CF_3)_2CHOH$ (a_P = 67.5 G, g = 2.0021).

On heating to 420 K, the *hfc* constants with the phosphorus nucleus for these radicals decrease to 64.6 G.

This indicates that the equilibrium between the complexes and free phosphorylfullerenyl radicals shifts on heating toward the free radicals:

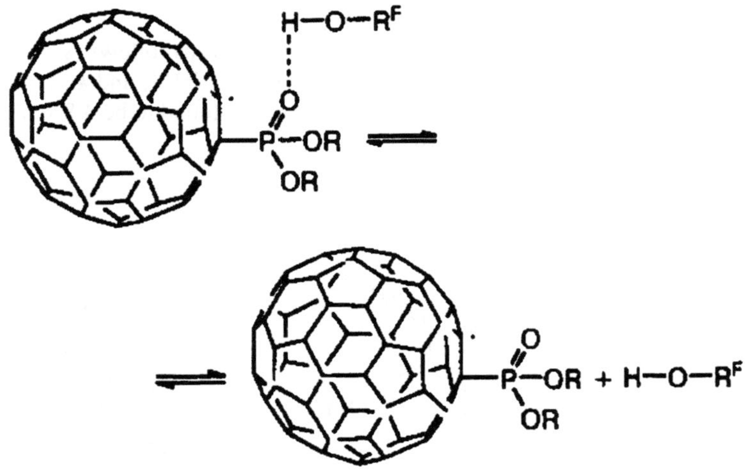

The EPR spectrum of $PF_3(O^tBu)C_{60}$ is also described.[54] In the trigonal bipyramidal arrangement of bonds around the phosphorus atom, two of the fluorine atoms occupy axial positions, as might be expected from their electronegativity. The C_{60} and O^tBu ligands and the third fluorine atom occupy equatorial positions.

1.5.2 Multiple addition of Phosphoryl Radicals

Multiple addition of phosphoryl radicals $^\bullet P(O)(OPr^i)_2$ to C_{60} at a fifteen-fold molar excess of a diphosphorylmercury compound with respect to fullerene was studied.[10]

Ultraviolet irradiation of this solution in the resonator of an EPR spectrometer results in six types of radicals (Fig. 20) characterized by the following constants of *hfc* of the unpaired electron with the ^{31}P nucleus and the following *g*-factors: a_P= 73.5 G, g = 2.0019; a_P= 66.75 G, g = 2.0023; a_P = 64.0 G, g = 2.0023; a_P = 4.25 G, a_P = 54.9 G, g = 2.0025; a_P = 33.5 G, g = 2.0023; a_P = 32.5 G, g = 2.0023.

The sequence of reactions giving these radicals is presented below:

$$Hg[P(O)(OPr^i)_2]_2 \longrightarrow {}^\bullet HgP(O)(OPr^i)_2 + {}^\bullet P(O)(OPr^i)_2$$

$$C_{60} + {}^{\bullet}P(O)(OPr^i)_2 \longrightarrow {}^{\bullet}C_{60}P(O)(OPr^i)_2$$

$${}^{\bullet}C_{60}P(O)(OPr^i)_2 + {}^{\bullet}P(O)(OPr^i)_2 \longrightarrow C_{60}[P(O)(OPr^i)_2]_2$$

$$C_{60}[P(O)(OPr^i)_2]_2 + {}^{\bullet}P(O)(OPr^i)_2 \longrightarrow {}^{\bullet}C_{60}[P(O)(OPr^i)_2]_3$$

$${}^{\bullet}C_{60}[P(O)(OPr^i)_2]_3 + {}^{\bullet}P(O)(OPr^i)_2 \longrightarrow C_{60}[P(O)(OPr^i)_2]_4$$

$$C_{60}[P(O)(OPr^i)_2]_n + {}^{\bullet}P(O)(OPr^i)_2 \longrightarrow {}^{\bullet}C_{60}[P(O)(OPr^i)_2]_{n+1}$$

n = 1, 3, 5 *etc.*

Since phosphoryl radicals, unlike benzyl radicals, do not form allyl radicals for steric reasons, the second phosphoryl radical recombines with the monoadduct at positions 1 and 4, while the third one attaches itself at a remote position.

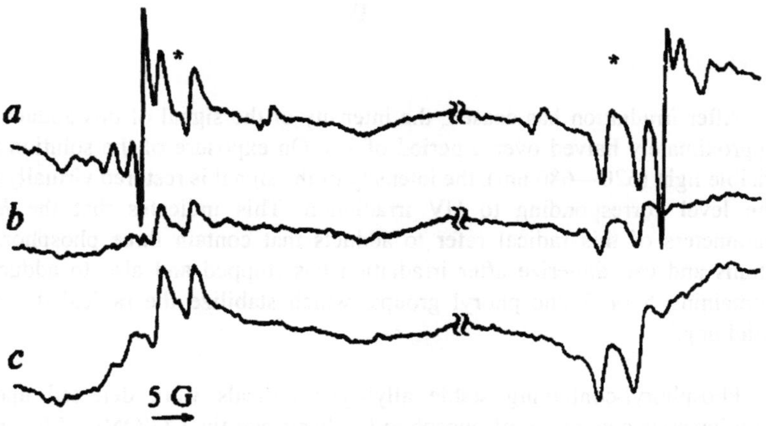

Figure 20. EPR spectra of the products of multiple addition of phosphoryl radicals to C_{60} at 293 K upon photolysis for 16 (*a*) and 32 (*b*) min and after termination of 48 min irradiation (*c*).

This gives rise to a fullerenyl radical (Fig. 20, *) in which two phosphoryl groups are non-equivalent relative to carbon bearing the maximum spin density. It is this radical that is responsible for the EPR spectrum which exhibits an additional splitting due to one phosphorus atom:[10]

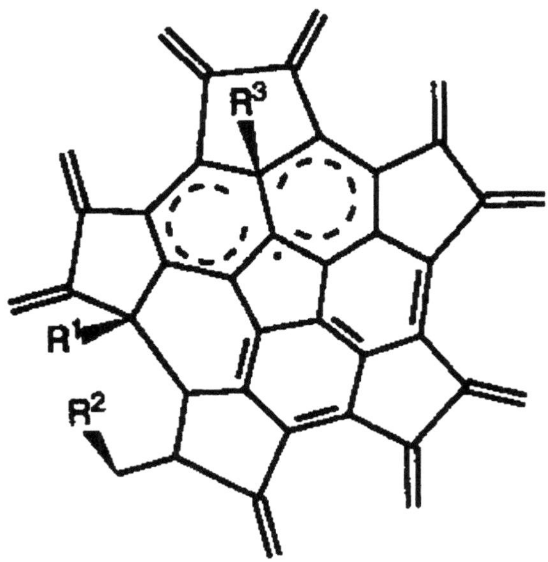

After irradiation has ceased, the intensity of the signal of this adduct is approximately halved over a period of 1 s. On exposure of the solution to visible light (620—680 nm), the intensity of the signal is restored virtually to the level corresponding to UV irradiation. This indicates that the *hfc* parameters of this radical refer to adducts that contain three phosphorus atoms and can dimerize after irradiation has stopped and also to adducts containing 5 or 7 phosphoryl groups, which stabilize the radical due to shielding.[10]

Phosphoryl-containing stable allyl-type radicals were detected upon simultaneous generation of phosphoryl radicals and the CO(OMe), Me, and CCl_3 radicals.[55] Their structure depends on the donor-acceptor properties of alkyl substituents. Simultaneous photochemical generation of phosphoryl and methoxycarbonyl radicals from the corresponding mercury compounds in a saturated toluene solution of C_{60} leads to both allyl type radicals containing a phosphoryl group in the center of the allyl system (a_P = 6.25 G, g = 2.0028) and those containing this group in a terminal position (a_P = 43.0 G, g = 2.0027):

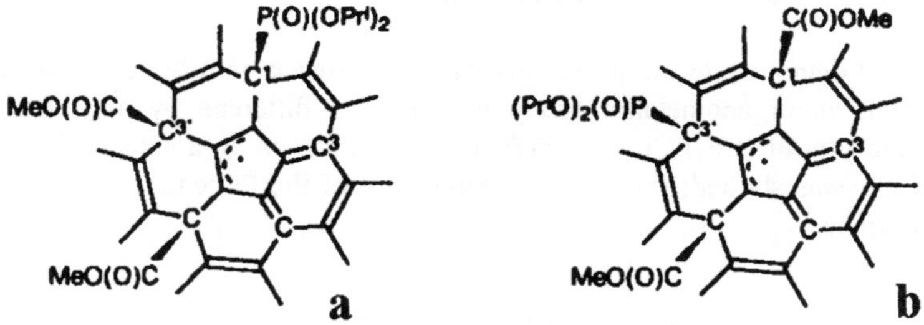

Simultaneous generation of phosphoryl and methyl radicals affords only allylic radicals containing the phosphoryl group in the center of the allyl system, the *hfc* constant at the phosphorus nucleus being 6.75 G:

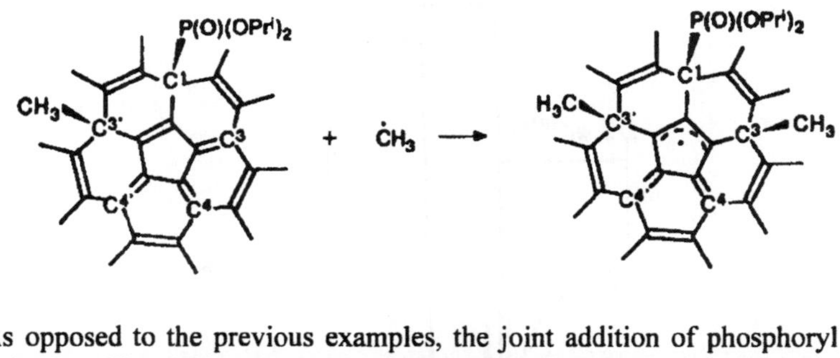

As opposed to the previous examples, the joint addition of phosphoryl and electrophilic trichloromethyl radicals yields allylic radicals containing two phosphoryl groups, which are located at the ends of the allyl system. This radical is responsible for an *hfc* constant with two equivalent ^{31}P nuclei (a_P = 41.25 G).

1.6 Addition of Metal-Centered Radicals

For organometallic fullerene chemistry, synthesis of the metal-containing fullerenyl radicals and their reactivity are of great interest. Catalytic activity of the stable palladium-fullerene complexes has been demonstrated previously and it is possible to expect catalytic properties of some kind from the metal-fullerenyl radicals. Combination of the fullerene core, transition metal, and free-radical center within one molecule can be done in different ways.

1.6.1 Addition of Pt-centered Radicals

To demonstrate the possibility of the existence of fullerenyl radicals in which an organometallic group is linked to fullerene by a σ-bond, the photolysis of $(CF_3)_2CF–Hg–Pt(PPh_3)_2–CH=CPh_2$-*cis* in a saturated solution of C_{60} was studied; this gave a spin-adduct of the Pt-centered radical with fullerene.[56]

$$(CF_3)_2CF–Hg–Pt(PPh_3)_2–CH=CPh_2 \xrightarrow{h\nu} (CF_3)_2CF–Hg^{\bullet}+{}^{\bullet}Pt(PPh_3)_2–CH=CPh_2$$

$$C_{60} + {}^{\bullet}Pt(PPh_3)_2–CH=CPh_2 \longrightarrow {}^{\bullet}C_{60}Pt(PPh_3)_2–CH=CPh_2$$

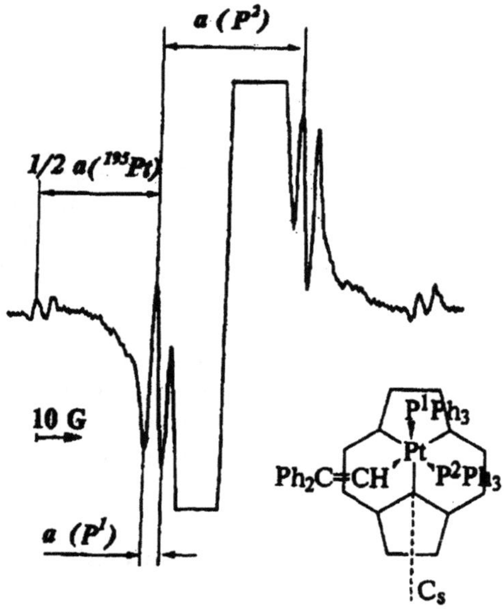

Figure 21. EPR spectrum of the radical ${}^{\bullet}C_{60}Pt(PPh_3)_2–CH=CPh_2$.[56]

It can be seen from Fig. 21 that this radical adduct is characterized by the following *hfc* constants: *a* (^{195}Pt) = 52.0 G, *a* (P^1) = 3.5 G, and *a*(P^2) = 30.5 G. The *hfc* constants with the ^{31}P nuclei are non-equivalent because of the non-symmetrical arrangement of the triphenylphosphine ligands located in the *cis*-positions with respect to the symmetry axis of the radical, one PPh_3 group being located above the five-membered ring (a_P = 30.5 G) and the other group being located above the six-membered ring (*a* = 3.5 G). This difference in the *hfc* constants, depending on the position of the substituent in the fullerenyl radical, is a typical feature of these radicals.

In addition to the radical adduct described, a broad line is recorded in the center of the spectrum, which is apparently related to the products of multiple addition of the carbon-centered radicals resulting from the homolytic cleavage of Hg—$CF(CF_3)_3$.

1.6.2 Addition of Re- and Sn-centered Radicals

The reaction of C_{60} with Bu_3SnH in toluene solution has been investigated. Mass spectra data indicate the formation of $C_{60}(BuH)_n$, (n = 1-3).[57]

The radical $^{\bullet}Re(CO)_5$ is considered to add to C_{60} to produce $C_{60}[Re(CO)_5]_2$.[58] The reaction has been monitored by IR. Photogenerated $C_{60}[Re(CO)_5]_2$ is unstable at room temperature in solution and has not been isolated.

1.7 Addition of Atoms

1.7.1 Addition of Hydrogen

1.7.1.1 Monoaddition

A topic of interest to $^{\bullet}C_{60}H$ has been connected with the extent of delocalization of the unpaired electron of the simplest mono-adducts.

To generate $^{\bullet}C_{60}H$, different methods were used:[59,60]

1. Photolysis of the reaction mixture consisting of thiophenol, tri-*n*-butyl tin hydride or 1,4 cyclohexadiene and C_{60}-saturated benzene;
2. Photolysis of di-*t*-butylperoxide, acetophenone, benzophenone, or acetone in the presence of various alcohols to produce radicals via photoreduction;
3. $^{\bullet}C_{60}H$ was also stabilized in solid neon, using microwave irradiation of H_2 and C_{60}.

The EPR spectrum of the radical adduct of hydrogen atom with C_{60} is characterized by the constant of the *hfi* with proton of 33 G. This value decreases with increasing temperature, according to the equation $a(T) = 94.44 - 0.00488T$ (MHz), which is valid in the 320 – 440 K temperature range.[59]

An interesting feature of the EPR spectrum of $^{\bullet}C_{60}H$ is that the low-field component of the doublet has a phase corresponding to emission, while the high-field component corresponds to absorption. This spectral pattern is an example of manifestation of chemically induced electron polarization (Fig. 22).

In the photochemical methods for the preparation of $^{\bullet}C_{60}H$, the principal stage is the photoreduction of C_{60} to anion radical $C_{60}^{-\bullet}$ followed by its protonation. Reducing agents in these reactions are different O-centered radicals, which can be generated via hydrogen atom abstraction from alcohols by *tert*-butoxyl radical or by photoexcited carbonyl group of various ketones (the highest intensity of $^{\bullet}C_{60}H$ was obtained with benzophenone).[60]

The formation of $^{\bullet}C_{60}H$ is explained by a competition of the addition reaction

$$(CH_3)_2C^{\bullet}OH + C_{60} \rightarrow (CH_3)_2C(OH)C_{60}^{\bullet}$$

with electron transfer, as was found to occur in UV-radiation chemical experiments, followed by proton transfer

$$(CH_3)_2C^{\bullet}OH + {}^{1,3}C_{60} \rightarrow [(CH_3)_2C^{+}OH + C_{60}^{-}] \rightarrow (CH_3)_2CO + {}^{\bullet}C_{60}H$$

The characteristic signal was observed in EPR spectrum with g = 2.0013, which assigned to the triplet state $^3C_{60}$ (Fig. 23).[60]

For the investigation of the character of delocalization of the unpaired electron in $^{\bullet}C_{60}H$, the fullerene enriched with ^{13}C (34.4%) was used.[60] There are intense center line from $^{\bullet}C_{60}H$ free of ^{13}C nuclei and minor lines from the radical with negligible coupling with ^{13}C nuclei, the other lines show intensities which clearly reflect single, double and quadruple degeneracy: 9.18 G (1C), 8.71 G (2C), 6.33 G (2C), 5.83 (1C), 4.38 G (2C), 4.0 G (2C), 3.63G (2C), 2.47 G (2C), 2.41 G (2C), 1.83 G (2C), 1.61 G (2C), 1.54 G (2C), 1.23 (2C), 1.09G (2C), 0.99 (2C), 0.83 (4C), 0.68 G (2C), 0.43 G (4C), 0.29 G (2C).

The couplings of forty carbon nuclei are resolved, which illustrates that the electron spin density is distributed over a large part of the sphere. Comparing the data obtained with the *hfc* constants of the unpaired electron with ^{13}C obtained for alkyl- and phosphorylfullerenyl radicals, one can observe the absence of a pair of lines corresponding to a coupling constant in the range of 16 – 20 G.

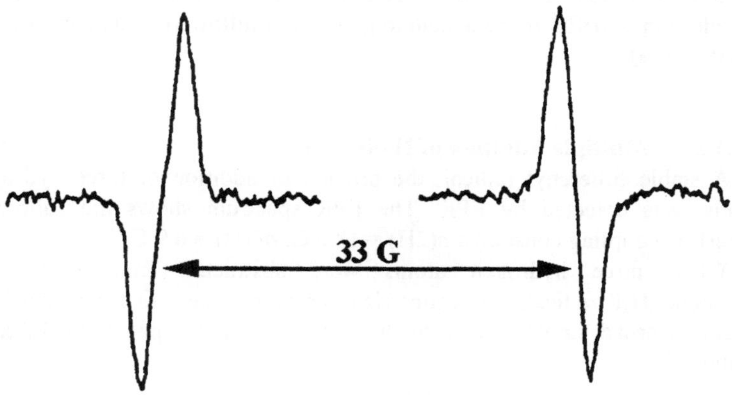

Figure 22. EPR spectrum of $^{\bullet}C_{60}H$. Adapted from ref 59.

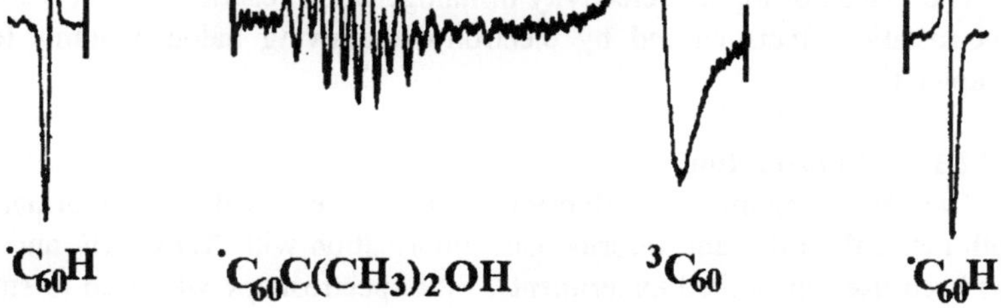

Figure 23. EPR spectra of $^{\bullet}C_{60}H$, $^{\bullet}C_{60}C(CH_3)_2OH$ and $^{3}C_{60}$. Adapted from ref 60.

The reactivity of $^{\bullet}C_{60}H$ was not specially studied, however, it is obvious that the radicals have to dimerize after irradiation terminates. Dimers HC_{60}–$C_{60}H$ were obtained preparatively.[61] The reaction was carried out in two steps. First, anion radical $C_{60}^{-\bullet}$ was synthesized, then it was protonated by the addition of trifluoroacetic acid to give $^{\bullet}C_{60}H$ followed by its dimerization (yield 4.9 %).

1.7.1.2 Multiple Addition of Hydrogens

A stable fullerenyl radical, the product of addition of three hydrogen atoms, was detected by EPR. The EPR spectrum shows the following hyperfine coupling constants: a(2H) = 18.7 G, a(1H) = 0.9 G.[28]

When three hydrogen atoms were abstracted from $C_{60}H_{36}$ by $^{\bullet}C(CF_3)_2C_6H_5F$ radical, no symmetrical structures seem to form, and EPR detects a broad singlet owing to the superposition of spectra of different isomers.[26]

1.7.2 Addition of Halogens

Fluorination, chlorination and bromination are the most important methods from the viewpoint of preparative synthesis through radical reactions. The halogen derivatives of fullerene are of particular interest on account of their application as synthons for preparing novel derivatives on their basis.[3]

The bromination and chlorination will add maximum of 24 halogens, whereas fluorination will readily add 48 fluorines.[3] The important distinction of halogenation reactions from other radical reactions is their regioselectivity and possibility to isolate derivatives containing different number of halogen atoms distributed in a regular manner over the surface of the carbon cage.

The reason of higher selectivity of halogenation reactions is likely to be electrostatic effects caused by electron-withdrawing halogen atoms (see Chapter 1).

1.7.2.1 Fluorination

Two main conditions of fluorination have been used, *viz.*, fluorination with metal fluorides and fluorine gas. Fluorination with XeF_2, BrF_5 and IF_5 has been used in limited experiments. EPR spectroscopy was used to study an adduct of one fluorine atom to C_{60} obtained by reaction with SF_5Cl.

Photolysis of a solution of C_{60} in toluene containing dissolved SF_5Cl at 260 K yielded the spectrum of $FC_{60}^{\bullet}$ described by the following parameters:[62] g = 2.00229, a (F) = 73.03 G. The followng mechanism was proposed:

$$SF_5 \xrightarrow{h\nu} {}^{\bullet}SF_5 + {}^{\bullet}Cl$$

$${}^{\bullet}SF_5 + C_{60} \longrightarrow {}^{\bullet}C_{60}SF_5$$

$${}^{\bullet}C_{60}SF_5 \longrightarrow SF_4 + {}^{\bullet}C_{60}F$$

The adduct of Cl to C_{60} (g = 2.00229, *a* (^{35}Cl) = 12.43 G) has been also obtained in this experiment.

Fluorination with metal fluorides upon heating in vacuum produces fluorofullerenes containing different number of fluorine atoms. Use of MnF_3 produces pure $C_{60}F_{36}$ consisting of two isomers, one of which has *T* symmetry, and the other has C_3 symmetry.[63] These isomers are obtained in 1 : 3 ratio, respectively. Their number and ratio parallel those found for hydrogenation which demonstrates the similarity between hydrogenation and fluorination, though these possesses are different in nature.

Fluorination of C_{60} with K_2PtF_6 produces $C_{60}F_{18}$.[64] The single-crystal X-ray crystallography confirms the conclusions that all fluorine atoms are bound to one hemisphere of C_{60} cage (Fig. 24).

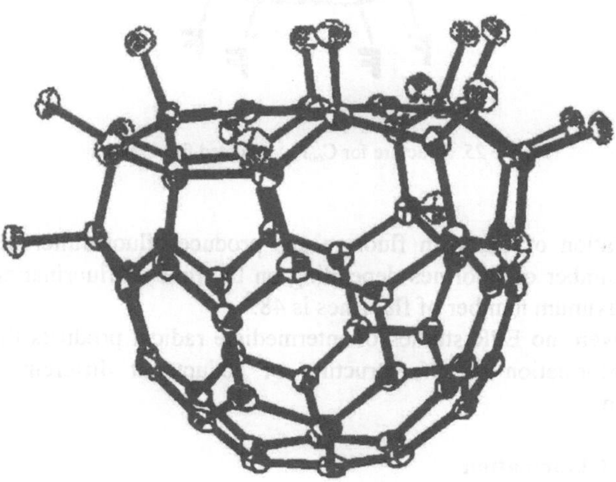

Figure 24. Structure for $C_{60}F_{18}$. Adapted from ref 64.

The use of Cs_2PbF_6 or KF/MnF_3 as fluorinating agent results in the formation of $C_{60}F_{20}$ whose ^{19}F NMR spectrum indicates that fluorine atoms are arranged in the equatorial positions of [60]fullerene. This compound was called "saturnene" (Fig. 25).[65]

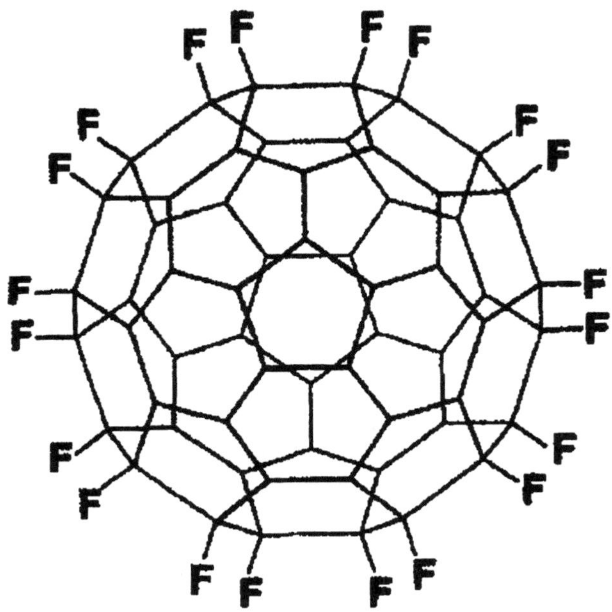

Figure 25. Structure for $C_{60}F_{20}$. Adapted from ref 65.

Fluorination of C_{60} with fluorine gas produces fluorofullerene with a different number of fluorines depending on the time of fluorination at 275 °C. The maximum number of fluorines is 48.[3]

There were no EPR studies of intermediate radical products that could provide information on the structure of adducts at different steps of fluorination.

1.7.2.2 Chlorination

Chlorination of C_{60} by chlorine gas at 250° gives an orange product indicated by chlorine uptake to be $C_{60}Cl_{24}$.[66]

The most important chloro compound yet made from C_{60} is $C_{60}Cl_6$ produced by reaction of C_{60} with ICl in benzene.[67] Its importance derives from the fact that it is soluble in many solvents and many derivatives may be made from it. The dechlorination of this compound under UV-irradiation has been studied by EPR (Fig. 26):[68]

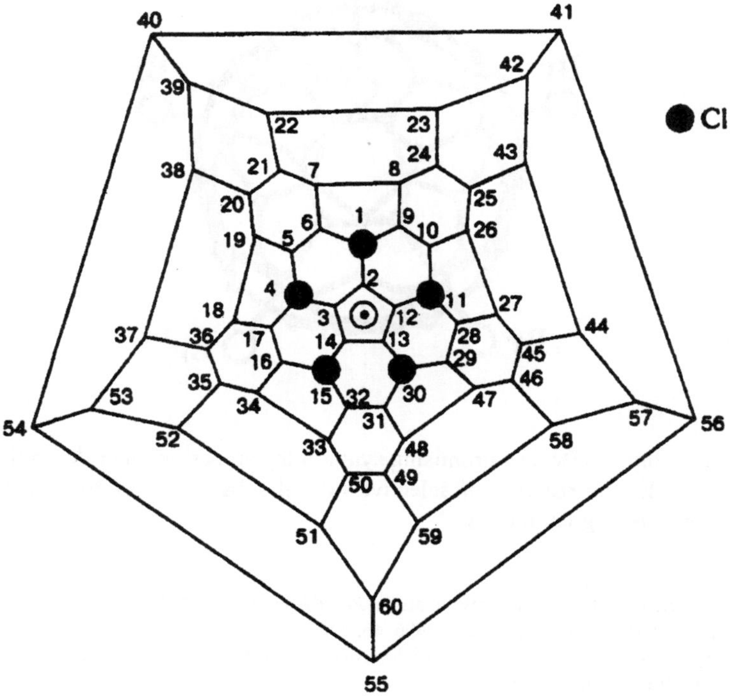

Figure 26. Structure for $^{\bullet}C_{60}Cl_5$.[68]

1.7.2.3 Bromination

The structures of bromofullerenes with different number of halogen atoms—$C_{60}Br_6$, $C_{60}Br_8$, $C_{60}Br_{24}$—were determined by X-ray diffraction.[69] The structure of $C_{60}Br_6$, analog of $C_{60}Cl_6$, is the most interesting.

The radical $^{\bullet}C_{60}Br_5$ is the precursor of the molecule $C_{60}Br_6$, whose structure was shown by X-ray crystallography to have five Br atoms around the central pentagon plus one Br atom on another carbon:

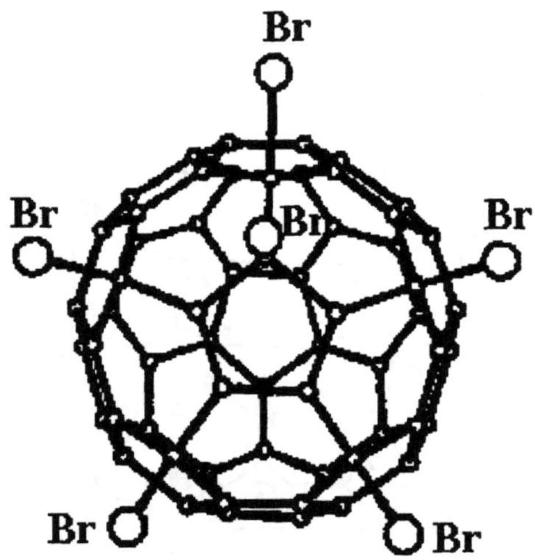

Compound $C_{60}Br_6$ is a promising synthon for preparing other derivatives. For example, the reaction of selective radical substitution of Br atoms by CF_3 groups is of great interest.[70]

References

1. J. R. Morton, F. Negri, K. F. Preston, *Magn. Res. Chem.*, 1995, **33**, 20.
2. B. L. Tumanskii, *Russ. Chem. Bull.*, 1996, **45**, 2267.
3. R. Taylor, *Lecture Notes on Fullerene Chemistry: A Handbook for Chemists*, Imperial College Press, London, 1999.
4. J. R. Morton, F. Negri, K. F. Preston, *Acc. Chem. Res.*, 1998, **31**, 63.
5. B. L. Tumanskii, V. V. Bashilov, O. G. Kalina, *Chemistry Reviews*, 2001, in press.
6. P. J. Krusic, E. Wasserman, B. A. Parkinson, B. Malone, E. R. Holler, P. N. Keizer, J. R. Morton, K. F. Preston, *J. Amer. Chem. Soc.*, 1991, **113**, 6274.
7. J. R. Morton, K. F. Preston, P. J. Krusic, S. A. Hill, E. Wasserman, *J. Phys. Chem.*, 1992, **96**, 3576.
8. J. R. Morton, K. F. Preston, P. J. Krusic, E. Wasserman, *J. Chem., Soc. Perkin Trans. 2*, 1992, 1425.
9. P. N. Keizer, J. R. Morton, K. F. Preston, P. J. Krusic, *J. Chem. Soc., Perkin Trans. 2*, 1993, 1041.
10. B. L. Tumanskii, V. V. Bashilov, E. N. Shaposhnikova, S. P. Solodovnikov, N. N. Bubnov, V. I. Sokolov, *Russ. Chem. Bull.*, 1996, **45**, 2538.
11. R. Borghi, L. Lunazzi, G. Placucci, P. J. Krusic, D. A. Dixon, N. Matsuzawa, M. Ata, *J. Amer. Chem. Soc.*, 1996, **118**, 7608.
12. G. A. Russell, K. Y. Chang, *J. Am. Chem. Soc.*, 1965, **87**, 4381.
13. I. V. Koptyug, A. G. Goloshevsky, I. S. Zavarine, N. J. Turro, P. J. Krusic, *J. Phys. Chem. A*, 2000, **104**, 5726.

14. W. T. Ford, T. Nishioka, F. Qiu, F. D'Souza, J. Choi, W. Kutner, K. Noworyta, *J. Org. Chem.*, 1999, **64**, 6257.
15. K. Ohkubo, S. Fukuzumi, *Inorg. React. Mechan.*, 2000, **2**, 147.
16. P. J. Fagan, P. J. Krusic, C. N. McEwen, J. Lazar, D. H. Parker, N. Herron, E. Wasserman, *Science*, 1993, **262**, 404.
17. J. R. Morton, K. F. Preston, *J. Phys. Chem.*, 1994, **98**, 4993.
18. J. R. Morton, F. Negri, K. F. Preston, G. Ruel, *J. Chem., Soc. Perkin Trans. 2*, 1995, 2141.
19. J. R. Morton, F. Negri, K. F. Preston, G. Ruel, *J. Phys. Chem.*, 1995, **99**, 10114.
20. M. Yoshida, D. Suzuki, M. Iyoda, *Chem. Lett.*, 1996, 1097.
21. B. L. Tumanskii, E. N. Shaposhnikova, V. V. Bashilov, S. P. Solodovnikov, N. N. Bubnov, S. R. Sterlin, *Russ. Chem. Bull.*, 1997, **46**, 1174.
22. M. Yoshida, F. Sultana, N. Uchiyama, T. Yamada, M. Iyoda, *Tetr. Lett.*, 1999, **40**, 735.
23. P. J. Fagan, P. J. Krusic, D. H. Evans, S. A. Lerke, E. Johnston, *J. Am. Chem. Soc.*, 1992, **114**, 9697.
24. a). N. M. Dimitrijevic, P. V. Kamat, R. W. Fessenden, *J. Phys. Chem.*, 1993, **97**, 615; b).R. G. Gasanov, O. G. Kalina, V. V. Bashilov, B. L. Tumanskii, *Russ. Chem. Bull.*, 1999, **48**, 2344.
25. P. J. Krusic, E. Wasserman, P. N. Keizer, J. R. Morton, K. F. Preston, *Science*, 1991, **254**, 1184.
26. B. L. Tumanskii, O. G. Kalina, V. V. Bashilov, A. V. Usatov, E. A. Shilova, Yu. I. Lyakhovetsky, S. P. Solodovnikov, N. N. Bubnov, Yu. N. Novikov, A. S. Lobach, V. I. Sokolov, *Russ. Chem. Bull.*, 1999, **48**, 1108.
27. B. L. Tumanskii, O. G. Kalina, V. V. Bashilov, *J. Mol. Mat.*, 2000, **13**, 23.
28. J. R. Morton, F. Negri, K. F. Preston, *Can. J. Chem.*, 1994, **72**, 776.
29. K. V. Scherer, T. Ono, K. Yamanouchi, R. Fernandez, F. Henderson, *J. Am. Chem. Soc.*, 1985, **107**, 718.
30. S. R. Sterlin, V. F. Cherstkov, B. L. Tumanskii, E. A. Avetisyan, *J. Fluor. Chem.*, 1996, **80**, 77.
31. a). Yu. I. Lyakhovetsky, E. A. Shilova, B. L. Tumanskii, A. V. Usatov, E. A. Avetisyan, S. R. Sterlin, A. P. Pleshkova, Yu. N. Novikov, Yu. S. Nekrasov, R. Taylor, *Full. Sci. Techn.*, 1999, **7**, 263; b). E. A. Shilova, Yu. I. Lyakhovetsky, A. I. Belokon', Yu. S. Nekrasov, *Mend. Commun.*, 1999, 176.
32. R. Borghi, L. Lunazzi, G. Placucci, G. Cerioni, A. Plumitallo, *J. Org. Chem*, 1996, **61**, 3327.
33. M. A. Cremonini, L. Lunazzi, G. Placucci, *J. Org. Chem.*, 1993, **58**, 4735.
34. Y.-K. Zhang, E. G. Janzen, Y. Kotake, *J. Chem., Soc. Perkin Trans. 2*, 1996, 1191.
35. T. Kusukawa, W. Ando, *Organometallics*, 1997, **16**, 4027.
36. B. L. Tumanskii, V. V. Bashilov, S. P. Solodovnikov, N. N. Bubnov, V. I. Sokolov, V. Ts. Kampel, A. Varshavskii, *Russ. Chem. Bull.*, 1994, **43**, 624.
37. B. L. Tumanskii, V. V. Bashilov, N. N. Bubnov, S. P. Solodovnikov, V. I. Sokolov, *Russ. Chem. Bull.*, 1995, **44**, 1771.
38. J. A. Baban, P. J. Marti, B. P. Roberts, *J. Chem., Soc. Perkin Trans. 2*, 1985, 1723.
39. B. L. Tumanskii, P. M. Valetsky, Yu. A. Kabachii, N. N. Bubnov, S. P. Solodovnikov, V. V. Korshak, A. I. Prokof'ev, M. I. Kabachnik, *Bull. Acad. Sci. USSR, Div. Chem. Sci.*, 1984, **33**, 2210.
40. E. A. Shilova, Yu. I. Lyakhovetsky, B. L. Tumanskii, V. V. Bashilov, Yu. S. Nekrasov, *School-Conference for Young Scientists "Organometallic Chemistry Towards the 21st Century"*, Moscow, 1999, 56.
41. B. W. Fullam, S. P. Mishra, M. C. R. Symons, *J. Chem. Soc, Dalt. Trans.*, 1974, 2145.

42. A. G. Davies, D. Griller, B. P. Roberts, *J. Am. Chem. Soc.*, 1972, **94**, 1782.
43. W. G. Bentrude, *Phosphorus and Sulfur*, 1977, **3**, 109.
44. B. L. Tumanskii, T. E. Kron, S. P. Solodovnikov, N. N. Bubnov, M. I. Kabachnik, *Izv. Akad. Nauk SSSR, Ser. Khim.*, 1982, 2692.
45. R. N. Nasirov, B. L. Tumanskii, N. A. Malysheva, N. A. Kardanov, N. N. Godovikov, N. N. Budnov, A. I. Prokof'ev, S. P. Solodovnikov, M. I. Kabachnik, *Bull. Acad. Sci. USSR, Div. Chem. Sci.*, **1985**, **34**, 383.
46. B. L. Tumanskii, R. Nasirov, S. P. Solodovnikov, N. N. Bubnov, N. A. Kardanov, R. G. Petrova, T. D. Churkina, I. I. Kandror, *Bull. Acad. Sci. USSR, Div. Chem. Sci.*, **1984**, **33**, 2028.
47. B. L. Tumanskii, T. V. Timofeeva, A. A. Kadyrov, N. N. Bubnov, S. P. Solodovnikov, K. N. Makarov, Yu. T. Struchkov, M. I. Kabachnik, *Dokl. Akad. Nauk SSSR*, 1990, **312**, 1152.
48. I. G. Neil, B. P. Roberts, *J. Organomet, Chem.*, 1975, **102**, C 17.
49. B. L. Tumanskii, O. G. Kalina, V. V. Bashilov, unpublished results.
50. D. L. Venezky, R. B. Fox, *J. Am. Chem. Soc.*, 1956, **78**, 1664.
51. B. L. Tumanskii, V. V. Bashilov, S. P. Solodovnikov, V. I. Sokolov, *Bull. Russ. Acad. Sci., Div. Chem. Sci.*, 1992, **41**, 1140.
52. B. L. Tumanskii, V. V. Bashilov, N. N. Bubnov, S. P. Solodovnikov, A. A. Khodak, V. I. Sokolov, *Russ. Chem. Bull.*, 1994, **43**, 1582.
53. B. L. Tumanskii, V. V. Bashilov, E. N. Shaposhnikova, S. P. Solodovnikov, N. N. Bubnov, V. I. Sokolov, *Russ. Chem. Bull.*, 1998, **47**, 1823.
54. J. R. Morton, K. F. Preston, Chem. Phys. Lett., 1996, **255**, 15.
55. B. L. Tumanskii, V. V. Bashilov, N. N. Bubnov, S. P. Solodovnikov, V. I. Sokolov, *Russ. Chem. Bull.*, 1995, **44**, 2316.
56. B. L. Tumanskii, V. V. Bashilov, O. G. Kalina, V. I. Sokolov, *J. Organomet. Chem.*, 2000, **599**, 28.
57. Y. M. She, X. H. Guo, Z. Y. Liu, S. Y. Liu, F. X. Fu, M. X. Qian, *Chem. J. Chin. Univ.*, 1996, **17**, 303.
58. S. Zhang, T. L. Brown, Y. Du, J. R. Shapely, *J. Am. Chem. Soc.*, 1993, **115**, 6705.
59. J. R. Morton, K. F. Preston, P. J. Krusic, L. B. Knight, *Chem. Phys. Lett.*, 1993, **204**, 481.
60. R. Klemt, E. Roduner, H. Fischer, *Chem. Phys. Lett.*, 1994, **229**, 524.
61. T. Tanaka, K. Komatsu, *J. Chem. Soc., Perkin Trans. 1*, 1999, 1671.
62. J. R. Morton, K. F. Preston, F. Negri, *Chem. Phys. Lett.*, 1994, **221**, 59.
63. O. V. Boltalina, J. M. Street, R. Taylor, *J. Chem., Soc. Perkin Trans. 2*, 1998, 649.
64. I. S. Neretin, K. A. Lyssenko, M. Yu. Antipin, Yu. L. Slovokhotov, O. V. Boltalina, P. A. Troshin, A. Yu. Lukonin, L. N. Sidorov, R. Taylor, Angew. Chem. Int. Ed., 2000, **39**, 3273.
65. O. V. Boltalina, V. Yu. Markov, P. A. Troshin, A. D. Darwish, J. M. Street, R. Taylor, *Angew. Chem. Int. Ed.*, 2001, **40**, 787.
66. G. A. Olah, et all, *J. Am . Chem. Soc.*, 1991, **113**, 9385.
67. P. R. Birkett, A. G. Avent, A. D. Darwish, H. W. Kroto, R. Taylor, D. R. M. Walton, *J. Chem. Soc., Chem. Commun.*, 1993, 1230.
68. R. G. Gasanov, O. G. Kalina, A. A. Popov, P. A. Dorozhko, B. L. Tumanskii, *Russ. Chem. Bull.*, 2000, **49**, 753.
69. P. R. Birkett, P. B. Hitchcock, H. W. Kroto, R. Taylor, D. R. M. Walton, *Nature*, 1992, **357**, 479.
70. I. S. Uzkih, E. I. Dorozhkin, O. V. Boltalina, L. N. Sidorov, "*Fullerenes and Atomic Clusters*", 5th Biennial International Workshop in Russia, 2001, 174.

Chapter 3

Radical Addition to C_{70}

Unlike C_{60}, in which all the carbon atoms are equivalent, other fullerene molecules are characterized by a lower symmetry; therefore, addition can result in several isomers. However, the reactivities of carbon atoms are different and the number of possible isomers depends on many reasons.[1,2]

In the case of C_{70} consisting of five types of non-equivalent carbon atoms in ratio 10:10:20:20:10, the formation of five spin-adducts is expected (see Fig. 2, Chapter 1):[3]

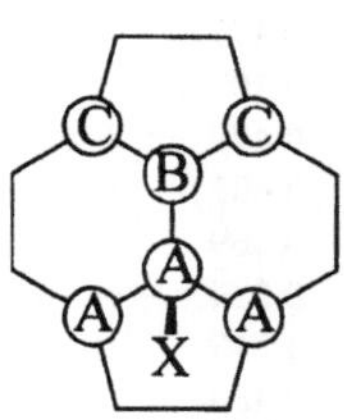

1

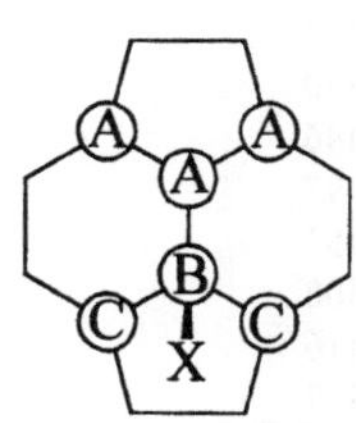

2

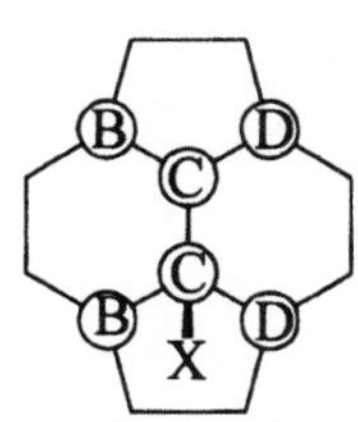

3

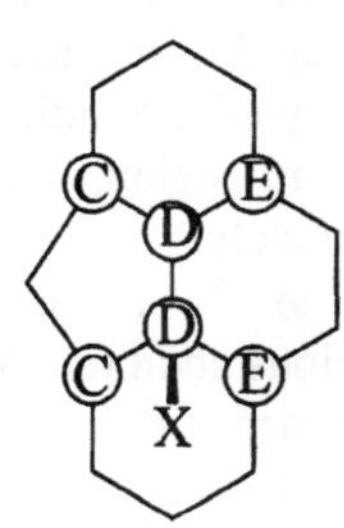

4

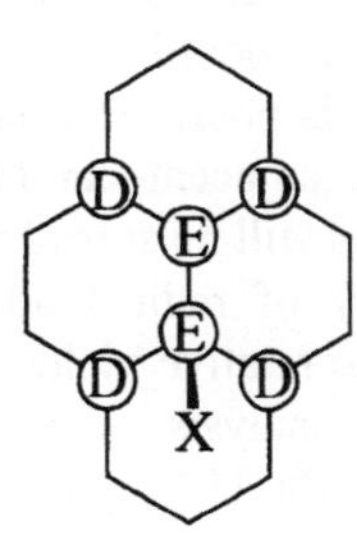

5

However, all five adducts were detected only upon the addition of methoxyl,[4] and trifluoromethyl radicals.[5] In other cases, a mixture of four or three isomeric spin-adducts is produced.[5]

Each type of radical species has been identified by comparing experimental data with the results of semiempirical calculations. These calculations indicate that of all the possible isomers, isomer E (Fig. 2, Chapter 1, and the scheme on p.85) resulting from the addition of a radical to carbon atom E in the fullerene cage is thermodynamically less stable than the isomers A–D formed upon attack by a radical on the corresponding carbon atoms; the most stable isomers are C and D (Table 1).[5,6]

Table 1. MNDO/PM3 and DFT/B3LYP calculations of the $HC_{70}^{\bullet}$ isomers. Adapted from ref 5.

Isomer	$\Delta(\Delta H)$, kcal mol^{-1}	a_H exp, G	a_H calc, G (normalized)
A	1.6	36.8	36.1
B	2.5	36.0	36.1
C	0.0	34.5	35.4
D	0.4	27.9	28.6
E	15.6		54.2

Kinetic effects, by contrast, lead to the expectation of preferential reactivity at the high order A–B and C–C bonds (Table 2).

Table 2. Bond lengths and orders in C_{70}.[1]

Bond	Bond length	π-bond orders
A-A	1.447	0.477
A-B	1.380	0.597
B-C	1.446	0.479
C-C	1.397	0.602
C-D	1.453	0.469
D-D	1.408	0.534
D-E	1.416	0.545
E-E	1.467	0.489

The kinetic effect requires that the regions of a fullerene where electron delocalization is considerable should be less susceptible to attack. Where delocalization in the product is poor, the spin density of the radical will be concentrated largely on the adjacent carbon, but in regions with higher aromaticity, the spin densities will be spread more widely.

Thus, the regiochemistry of radical addition to C_{70} is not yet well understood. For example, the relative yields in various additions follow the order: D>C>B>A>E. This analysis is based in part on the calculated thermodynamic stabilities of the $HC_{70}^{\bullet}$ radical, but the order is opposite to the observations for most other additions to fullerene[70] (*e.g.*

hydrogenation, cycloadditions, and addition of organometallic reagents) in which the positional reactivity order is A,B > C: (kinetic order); the only additions at the D, D positions occur in chlorination (1,4) and with benzyne (1,2) (where it is a minor component of the overall reaction).

Moreover, the spin densities are calculated to be most delocalized for addition to D and E atoms, since these are the most aromatic regions of the molecule. By contrast to addition to A, B, and C atoms, the spin density is localized mainly on the β-carbon. These latter radicals should therefore be the most stable, due to shielding by the addend. Since the *opposite* is true, then either the assignments are incorrect or one must seek an additional factor, which could be the greater curvature of the C_{70} surface at the poles.[7]

The DFT calculations of all possible isomers of $^{\bullet}C_{70}H$ were done to estimate the *hfc* constant and find out what lines in the $^{\bullet}C_{70}H$ EPR spectrum are relevant to the isomer (Table 1).[5]

The isotropic hyperfine interaction measured in the EPR experiment is proportional to the s character of the MO at a particular nucleus. The DFT spin densities in 1s orbitals of the hydrogens in the five isomers of $HC_{70}^{\bullet}$ can be converted to isotropic hyperfine interactions by multiplication by 504 G - the hyperfine interaction of the isolated hydrogen atom with unit spin density. The values obtained in this manner are systematically too low and they are even lower using the results of semiempirical calculations. Relative to each other, however, they are quite satisfactory for the first four isomers of HC_{70}.[5]

1.1 Addition of Carbon-Centered Radicals

The adducts $^{\bullet}C_{70}R$ exhibit three common features:

1. The g-factors for the spin-adducts of each type are different;
2. The constants of *hfc* with the key atom (i.e., the atom that carries the highest spin density) in the added radical are different;
3. One of the isomers can be stable in the absence of irradiation.

The EPR spectrum obtained by photolysis of di-*tert*-butyl peroxide in the presence of C_{70} and isobutan indicates addition of *tert*-butyl radicals to the C_{70}, forming the spin-adduct in three isomeric forms (g = 2.0027, 2.0024, 2.0021).[3]

Hyperfine interactions with the protons of the *tert*-butyl groups are clearly visible. The confirmation of the identity of these isomers as mono-adducts of C_{70} was provided by the use of isobutane enriched with ^{13}C at the central carbon atom (Fig. 1).

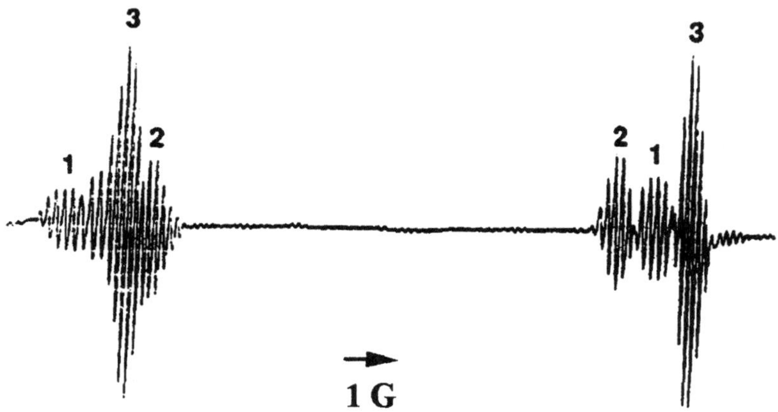

Figure 1. EPR spectrum of $Me_3{}^{13}CC_{70}^{\bullet}$ at 300 K. Adapted from ref 3.

A similar result was obtained by photolysis of C_{70} solutions in ^{13}C-enriched CCl_4. Three isomers of $CCl_3C_{60}^{\bullet}$ were observed, having ^{13}C *hfi* of 34.6, 30.5, and 26.7 G.[3]

While simple alkyl radicals afford only three regioisomers, the more reactive aryl and fluoroalkyl radicals give rise to four, except for the trifluoromethyl radical which yielded the EPR spectra of all five expected isomers.[5]

The spectra of the CF_3-$C_{70}^{\bullet}$ isomers are all quartets for three equivalent fluorines above 285 K indicating that, at these temperatures, the CF_3 group in all four regioisomers is rotating freely about the bond connecting it to the C_{70} surface. Upon shuttering the spectra at 320 K, the isomers (2), (3), and (4) (Table 3) were not detected, while the spectrum of (1) grows in intensity, and a new weaker quartet spectrum appears with the relatively high g-factor of 2.0028 and a CF_3 fluorine splitting of 0.33 G (Fig. 2).[5,6]

Table 3. EPR parameters of the radical adducts $CF_3C_{70}^{\bullet}$.[5]

isomer	a_F, G	g-factor	T, K
1	0.05	2.00251	320
2	0.25	2.00271	320
3	0.13	2.00218	320
4	0.13	2.00283	320
1	0.035	2.00251	360
2	0.273	2.00271	360
3	0.100	2.00218	360

isomer	a_F, G	g-factor	T, K
4	0.165	2.00283	360
5	0.330	2.00208	360

If the temperature is raised with no further irradiation to 360 K, the spectra of isomers (2), (3) and (4) (Table 3) grow to afford a composite spectrum of fife CF_3 fluorine quartets. The persistency of (1) and (5) at relatively low temperatures and in the absence of UV irradiation indicates that isomers of this type are intrinsically less prone to dimerization than those of type (2), (3) and (4).

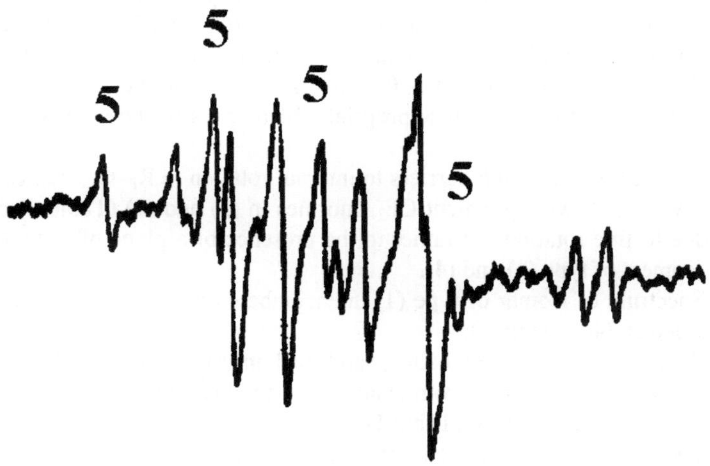

Figure 2. EPR spectrum of $CF_3C_{70}^{\bullet}$ at 360 K after UV irradiation. Adapted from ref 5.

The more complicated composite spectrum of isomeric adducts CF_3CF_2-$C_{70}^{\bullet}$ is again the superposition of the spectra of four isomers as can be verified by computer simulation (Table 4).[5,6]

Table 4. EPR parameters of radical adducts $CF_3CF_2C_{70}^{\bullet}$.[5]

isomer	a_F, G	g-factor	T, K
1	0.20 (1F), 0.04 (1F), 1.35 (3F)	2.00240	380
2	0.58 (1F), 0.63 (1F), 2.13 (3F)	2.00260	380
3	0.125 (2F), 2.27 (3F)	2.00202	380
4	0.53 (2F), 2.52 (3F)	2.00273	380

The spectra of isomers (3) and (4) (Table 4) have the same splitting pattern as the C_{60} analog. Isomers (1) and (2) are different.

In (1) only one of the two CF_2 fluorines interacts appreciably while the second has a splitting that can barely be resolved, while in (2) the two CF_2 fluorines are equivalent at 450 K but not at 380 K.

The observation of two distinct CF_2 fluorines in CF_3CF_2-C_{70} (1) and (2) implies these two radicals, and consequently all homologous isomers (1) and (2) (Table 4), to be without a plane of symmetry and the fluoroethyl group to be not free to rotate around the bond connecting it to the C_{70} surface even at temperatures as high as 470 K.[5]

The latter agrees with behavior of fluoroalkyl radical adducts of C_{60} higher than the $CF_3^{\bullet}$ adduct—such as $CF_3CF_2^{\bullet}$, $(CF_3)_2CF^{\bullet}$, and $(CF_3)_3C^{\bullet}$—in which the fluoroalkyl substituent is essentially locked in the staggered conformation relative to the C_{60} framework below it on the time scale of the EPR experiment (< 1 μs) as appropriate for barriers of the order of 10 kcal mol^{-1}.[5]

Because of these high barriers to internal rotation in R_F—$C_{60}^{\bullet}$ radicals, the observation of two equivalent CF_2 fluorines in (3) and (4) (Table 4) cannot be due to free rotation but rather to the presence of a plane of symmetry in the isomers of type (3) and (4).

Spectrum of isomer of type (1) unencumbered by the other isomers was recorded at 340 K (Fig. 3).[5]

Photoreactions of aryl halides and aryl mercury compounds with C_{70} yielded composite spectra from which the individual spectra of four isomeric species could be dissected (Table 5).[5]

The correctness of this analysis was verified by computer simulations. The splitting patterns are the same as those of the C_{60} analogs differing only slightly in the magnitudes of the splittings. Each isomer gives rise to a triplet for the two *meta* protons.

The spectrum obtained in the photoreaction of C_{70} with diphenylmercury in benzene is shown in Figure 4 together with the computer simulation based on the parameters for four isomers.

Analogous composite spectra were obtained when C_{70} was reacted with several other photochemically generated radicals and were similarly analyzed by computer simulation. In each case the spectra were due to four isomeric adducts whose EPR parameters are listed in Table 6.

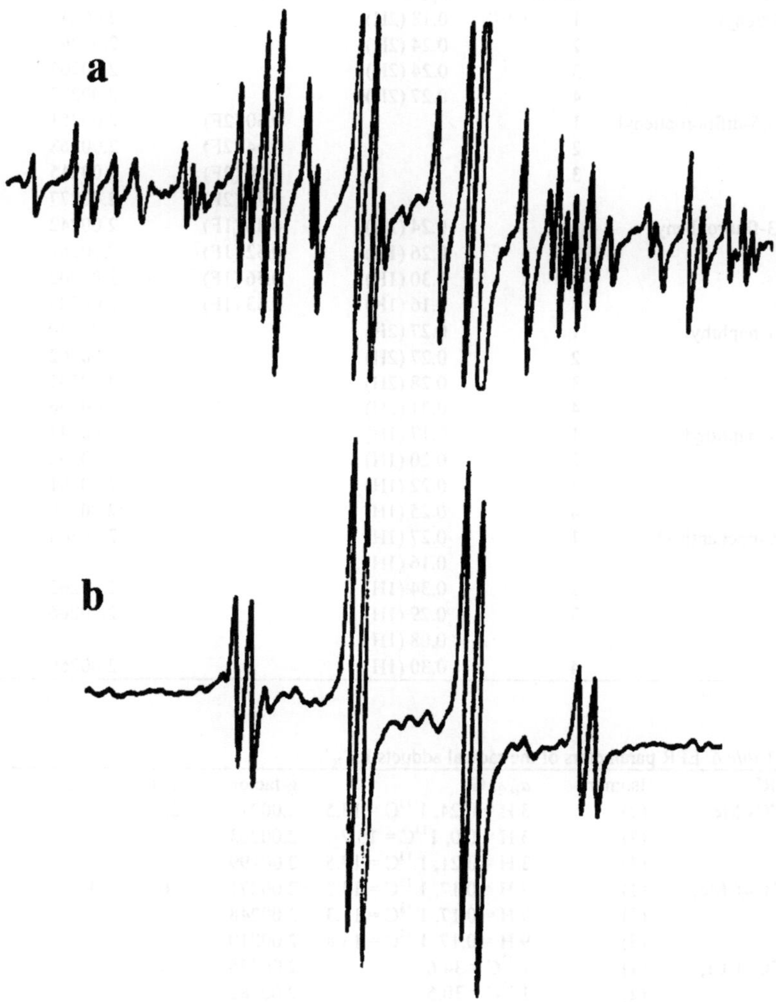

Figure 3. EPR spectra of $CF_3CF_2C_{70}^{\bullet}$ (a) under UV irradiation at 380 K and (b) without irradiation. Adapted from ref 5.

Table 5. EPR parameters of the radical adducts $Ar\text{-}C_{70}^{\bullet}$.[5]

R	isomer	a_H, G	a_F, G	g-factor
Phenyl	1	0.18 (2H)		2.00248
	2	0.24 (2H)		2.00266
	3	0.24 (2H)		2.00207
	4	0.27 (2H)		2.00272
3,5-difluorophenyl	1		0.60 (2F)	2.00251
	2		0.64 (2F)	2.00268
	3		0.78 (2F)	2.00215
	4		0.71 (2F)	2.00277
3-fluorophenyl	1	0.24 (1H)	0.46 (1F)	2.00242
	2	0.26 (1H)	0.52 (1F)	2.00261
	3	0.30 (1H)	0.56 (1F)	2.00202
	4	0.16 (1H)	0.33 (1F)	2.00283
1-naphthyl	1	0.27 (2H)		2.00240
	2	0.27 (2H)		2.00262
	3	0.28 (2H)		2.00201
	4	0.31 (2H)		2.00266
2-naphthyl	1	0.17 (1H)		2.00243
	2	0.20 (1H)		2.00262
	3	0.22 (1H)		2.00201
	4	0.25 (1H)		2.00269
9-phenanthryl	1	0.27 (1H)		2.00244
		0.16 (1H)		
	2	0.34 (1H)		2.00262
	3	0.29 (1H)		2.00206
		0.08 (1H)		
	4	0.30 (1H)		2.00269

Table 6. EPR parameters of the radical adducts $RC_{70}^{\bullet}$.

$R^{\bullet}$	Isomers	$a_{H,C}$, G	g-factor	T, K	Ref.
$^{\bullet}C_{70}Me$	(1)	3 H = 0.24, 1 ^{13}C = 13.5	2.0021	280	6
	(2)	3 H = 0.0, 1 ^{13}C = 16.7	2.00263		
	(3)	3 H = 0.21, 1 ^{13}C = 17.5	2.00199		
$^{\bullet}C_{70}CMe_3$	(1)	9 H = 0.17, 1 ^{13}C = 14.2	2.00271	300	3
	(2)	9 H = 0.17, 1 ^{13}C = 11.3	2.00248		
	(3)	9 H = 0.17, 1 ^{13}C = 13.8	2.00210		
$^{\bullet}C_{70}CCl_3$	(1)	1 ^{13}C = 34.6	2.00425	300	3
	(2)	1 ^{13}C = 30.5	2.00382		
	(3)	1 ^{13}C = 26.7	2.00345		

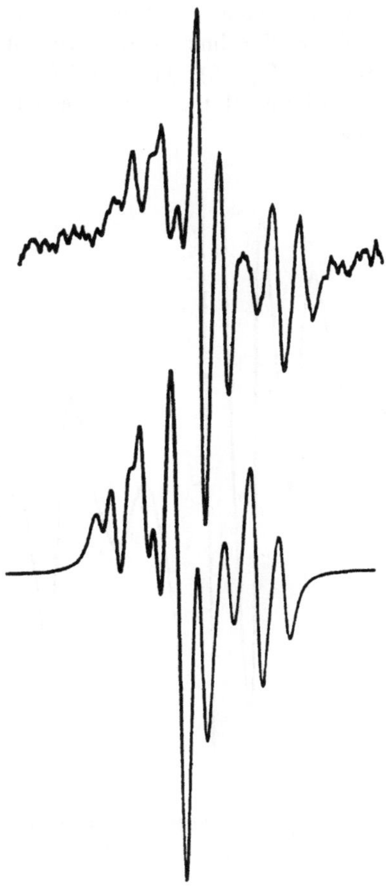

Figure 4. EPR spectrum of $^{\bullet}C_{70}C_6H_5$ and its simulation. Adapted from ref 6.

1.2 Addition of S- and O-centered Radicals

Three regioisomers of MeS–$C_{70}^{\bullet}$ adduct were obtained by photolysis of MeSSMe in the presence of C_{70}.[8] Each regioisomer exhibits a 1 : 3 : 3 : 1 quartet, due to the coupling of the unpaired electron with the three equivalent hydrogens of the CH_3S group. The corresponding *hfc* constants are similar to those reported for the MeS–$C_{60}^{\bullet}$ radical (Table 7).

Photolysis of di-*tert*-butyl peroxide in the presence of C_{70} also yields an EPR spectrum due to few isomeric t-BuO-$C_{70}^{\bullet}$ adducts. In the case of photolysis of MeOSSOMe, five lines of different intensity were observed due to MeO-$C_{70}^{\bullet}$ radical (Fig. 5). The MeO-$C_{70}^{\bullet}$ radical adducts thus represent an example where all the five expected regioisomers can be detected by EPR (Table 7).[8]

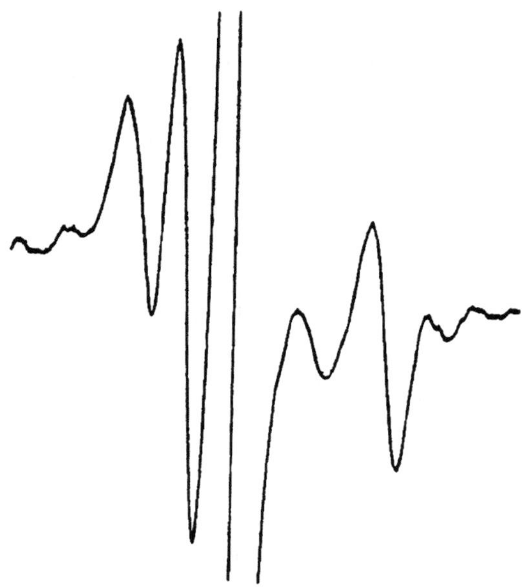

Figure 5. EPR spectrum of $CH_3OC_{70}^{\bullet}$. Adapted from ref 8.

Table 7. EPR parameters of the radical adducts $RC_{70}^{\bullet}$.[8]

$R^{\bullet}$	Isomer	a_{Me}, G	g - factor	Ref.
$MeSC_{70}$	(1)	0.39	2.00082	8
	(2)	0.47	2.00177	
	(3)	0.41	2.00188	
$MeOC_{70}$	(1)		2.00172	8
	(2)		2.00193	
	(3)		2.00216	
	(4)		2.00231	
	(5)		2.00245	

1.3 Addition of B-centered Radicals

Addition of a boron-centered carboranyl radical to C_{70} affords three isomers characterized by the following *hfc* constant with ^{11}B nucleus (Fig. 6): a (^{11}B) = 20.0 G, g = 2.0021; a (^{11}B) = 19.7 G, g = 2.0032; a (^{11}B) = 15.5 G, g = 2.0026; a (^{10}B) = 5.1 G.[9] The first two radical adducts decay within less than 1 s after the UV light is switched off, and one of the radical adducts has a lifetime of more than 10 min.

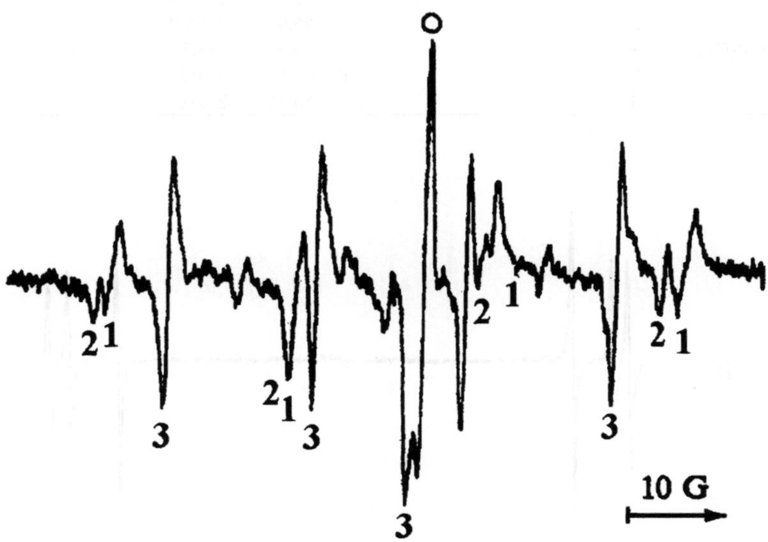

Figure 6. EPR spectrum of three isomers of $^{\bullet}C_{70}B_{10}H_9C_2H_2$-*m*.[9]

1.4 Addition of Phosphoryl Radicals

Addition of phosphorus-centered radicals to C_{70} is of particular interest because the large constant of hyperfine interaction with phosphorus nucleus provides an opportunity to discriminate isomers close in structure and to analyze multiple addition products by EPR.[2]

The formation of three isomers was detected for the addition of $^{\bullet}P(O)(OMe)_2$ produced both by hydrogen abstraction form appropriate hydrophosphoryl compound[3] and by photolysis of diphosphorylmercury compounds (Fig. 7) (Table 8).[10]

Table 8. EPR parameters of the radical adducts $^{\bullet}C_{70}P(O)(OR)_2$.

R	isomer	a_H, G	a_P, G	g-factor	Ref
$P(O)(OMe)_2$	1	6 H = 0.12	1 P = 71.2	2.00277	3
	2	6 H = 0.12	1 P = 66.8	2.00269	
	3	6 H = 0.08	1 P = 55.9	2.00248	
$P(O)(OEt)_2$	1		1 P = 70.0	2.0039	11
	2		1 P = 65.6	2.0033	
	3		1 P = 55.0	2.0029	
$P(O)(OPr^i)_2$	1		1 P = 70.5	2.0031	10
	2		1 P = 66.5	2.0027	
	3		1 P = 55.25	2.0028	

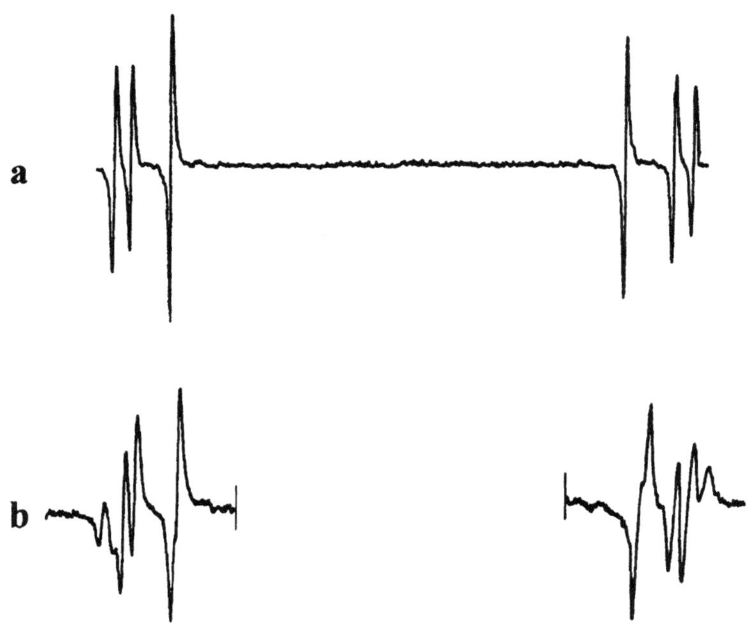

Figure 7. EPR spectra of 3 isomers of $^{\bullet}C_{70}P(O)(OPr^i)_2$ at (a) 290 K and (b) 200 K.[11]

After irradiation has been terminated, two radicals decay over less than one second, while the third one survives for several days. The intensity of this signal is rather high that allows one to observe several pairs of satellites due to coupling to ^{13}C nuclei of the fullerene scaffold: a_1 = 16.25 G, a_2 = 11.63 G, a_3 = 6.5 G, a_4 = 5.75 G, a_5 = 1.75 G, a_6 = 1.0 G (Fig. 8).[11]

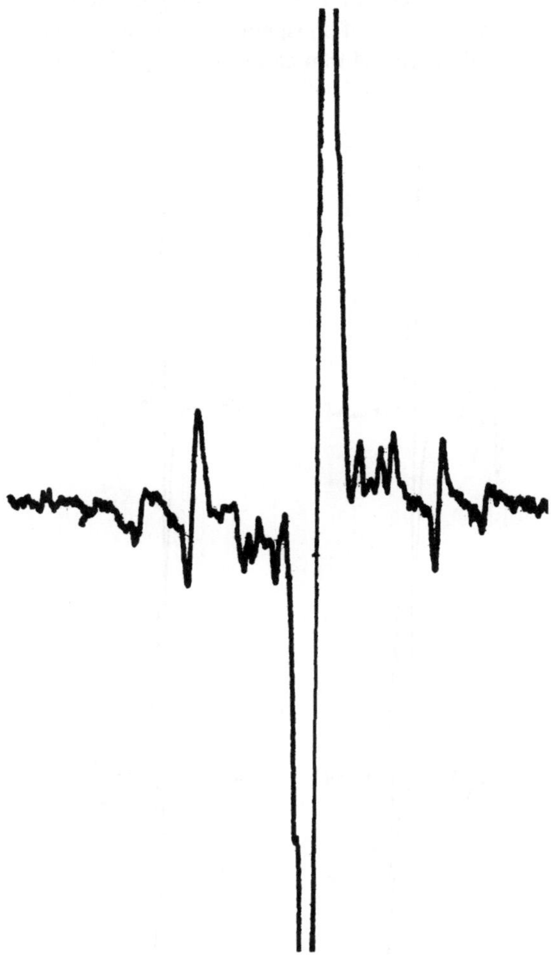

Figure 8. Low field component of the stable $^{\bullet}C_{70}P(O)(OPr^i)_2$ (table 8, isomer 3) regioisomer.[11]

The EPR spectra of the radical adducts were studied in the temperature range 190 – 430 K.[11] A line broadening due to hindered rotation of phosphoryl group between two conformations is also observed similarly to $^{\bullet}C_{60}P(O)(OR)_2$ as temperature decreases.

At 190 K, the EPR spectra of these "frozen" conformations have the following *hfc* constants: $a(^{31}P)$ = 63.7 G, g = 2.0029; $a(^{31}P)$ = 67.0 G, g = 2.0031; $a(^{31}P)$ = 68.1 G, g = 2.0034; $a(^{31}P)$ = 71.5 G, g = 2.0029 (Fig. 7b).

The shape of EPR spectra obtained by heating the adducts of phosphoryl radicals to C_{70} is independent of phosphoryl structure. In all cases, heating of $^{\bullet}C_{70}P(O)(OR)_2$ to 400 K for 24 min causes decomposition of the least stable isomer (Fig. 9).[11]

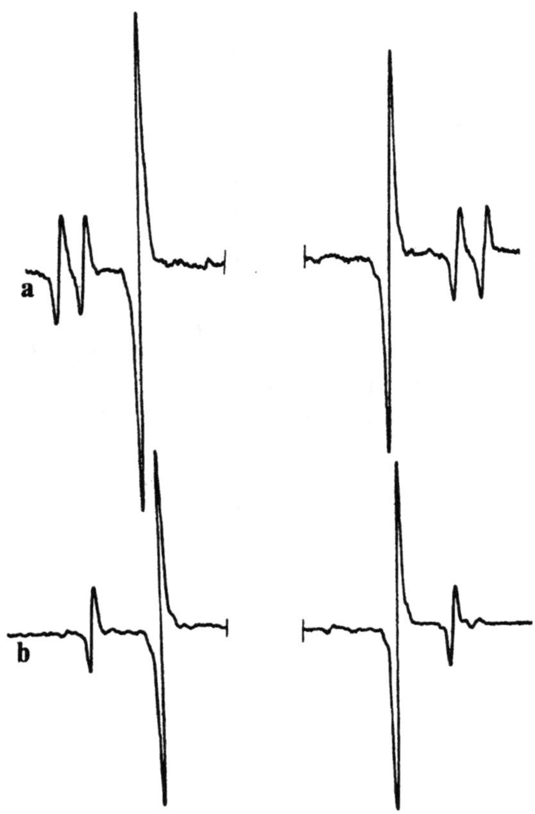

Figure 9. EPR spectra of $^{\bullet}C_{70}P(O)(OEt)_2$ at (a) 300 K and (b) 400 K.[11]

Multiple addition of phosphoryl radicals to C_{70} was also studied, which takes place on prolonged irradiation of a toluene solution containing an excess of C_{70} and diphosphorylmercury.[11,12]

In the case of $^{\bullet}P(O)(OPr^i)_2$, the EPR spectrum recorded after photolysis at ambient temperature for about an hour is a complex set of lines.

Then the sample was heated at 400 K to remove the least stable polyadducts or to isomerize them through migration of phosphoryl group to the most stable isomer.

The resultant EPR spectrum showed the presence of two fullerenyl radicals: an allyl-type radical and a product of addition to the apex D.

After addition of $(CF_3)_3OH$ to increase *hfc* constants with phosphorus nucleus, a spectrum similar to allyl radical $C_{70}H_3$ was obtained in which the terminal carbons of the allyl system were nonequivalent (Fig. 10) ($a\,(^{31}P_1) = 47.4$ G, $a\,(^{31}P_2) = 48.05$ G, $a\,(^{31}P_3) = 5.4$ G).[11]

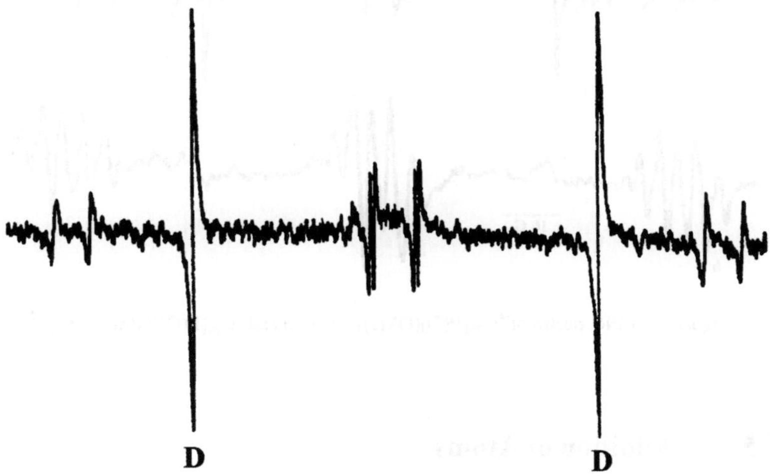

Figure 10. EPR spectrum of $^{\bullet}C_{70}[P(O)(OPr^i)_2]_3$ at 300 K in the presence of $(CF_3)_3OH$.[11]

Multiple addition of $^{\bullet}P(O)(OMe)_2$ radicals generated in a similar way was also studied.[12] After photolysis carried out for 0.5 h at 300 K, the EPR spectrum of an allyl-type radical was recorded; the *hfc* constant with two phosphorus nuclei is $a_P(2P) = 44.5$G and that with one phosphorus nucleus is $a_P(1P) = 5.0$G (Fig. 11*a*).

Further photolysis results in the attachment of two additional radicals. The values of the *hfc* constants, $a_P(2P)$ = 48,5 G, $a_P(1P)$ = 6,0 G, and $a_P(2P)$ = 3,25 G, imply that, unlike five benzyl radicals, which add to C_{60} to give a cyclopentadienyl radical, the allylic structure is retained in the case of $^{\bullet}C_{70}[P(O)(OMe)_2]_5$ and the additional *hfc* is due to the attachment of two phosphoryl radicals in the δ-position with respect to the terminal carbons of the allylic system (Fig. 11*b*).

The rates of addition of phosphoryl radicals to C_{70} were determined by competitive reaction technique. The values of rate constants for the radical addition to nonequivalent apices of the carbon polyhedron are 1.9, 2.2, and 2.4 10^{-8} l mol^{-1} s^{-1}.[13]

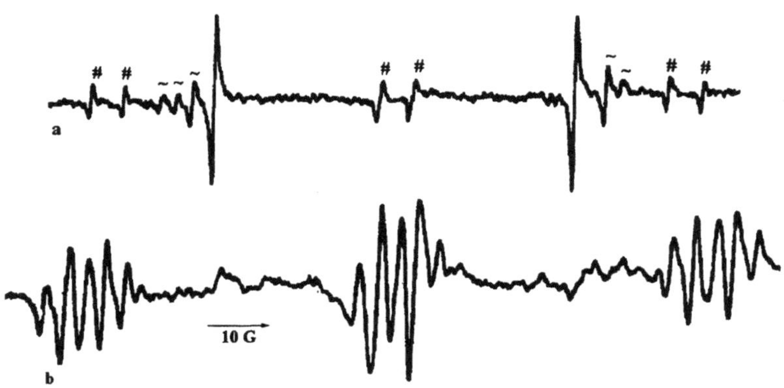

Figure 11. EPR spectra of $^{\bullet}C_{70}[P(O)(OMe)_2]_3$ (#) (a) and $^{\bullet}C_{70}[P(O)(OMe)_2]_5$ (b).[12]

1.5 Addition of Atoms

1.5.1 Addition of Hydrogen

When a benzene solution of C_{70} containing benzophenone and *iso*-propanol is exposed to radiation, hydrogens are generated, which add to C_{70} to give adducts of 5 different types, four of which are characterized by a splitting with one hydrogen atom, while the fifth adduct is responsible for *hfc* with three protons, indicating that several hydrogen atoms added and that an allylic radical formed. Chemically induced electron polarization is manifested in all the spectra (Fig. 12).[14]

Spin adducts of three types are detected when $C_{70}H$ is generated by photolysis of a saturated benzene solution containing 1,2-cyclohexadiene.[6]

The number of possible isomers of H_3C_{70} is 2792, so observation of only one of them is quite remarkable and the assignment of spectral line to a specific isomer of $H_3C_{70}^{\bullet}$ is not obvious.[14]

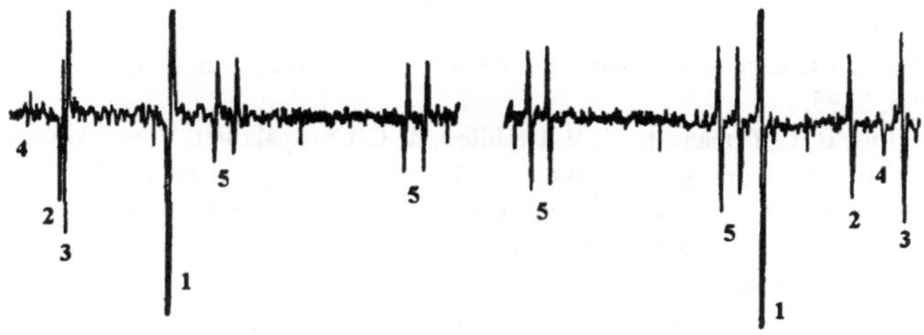

Figure 12. EPR spectra of $^{\bullet}C_{70}H$ (1-4) and $^{\bullet}C_{70}H_3$ (5). Adapted from ref 14.

1.5.2 Addition of Halogen Atoms

The addition of fluorine atoms, which were generated similarly to procedure used for fluorination of C_{60}, to C_{70} resulted in 4 isomers (g = 2.00148, a_F=53.54 Гс; g = 2.00181, a_F=65.25 Гс, g = 2.00187, a_F=74.21 Гс, g = 2.00133, a_F=74.46 Гс).[15]

Fluorination of C_{70} by MnF_3 gives a mixture of $C_{70}F_{36/38/40}$ with $C_{70}F_{38}$ as the main product.[16]

Chlorination of C_{70} with ICl leads to $C_{70}Cl_{10}$ where chlorine atoms are arranged as shown in diagram:[17]

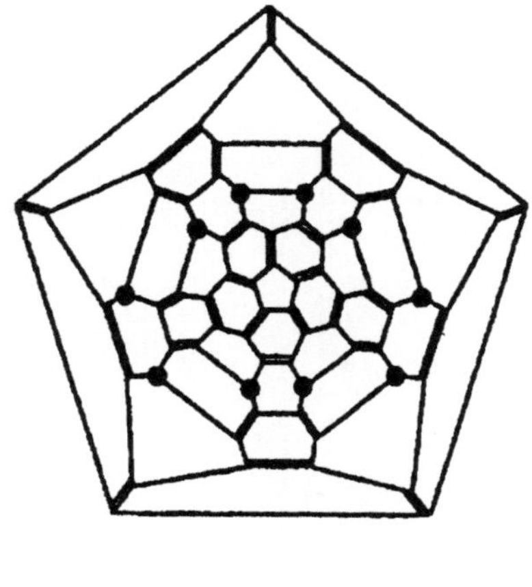

References

1. R. Taylor, *Lecture Notes on Fullerene Chemistry: A Handbook for Chemists*, Imperial College Press, London, 1999.
2. B. L. Tumanskii, V. V. Bashilov, O. G. Kalina, *Chemistry Reviews*, 2001, in press.
3. P. N. Keizer, J. R. Morton, K. F. Preston, *J. Chem. Soc., Chem. Commun.*, 1992, 1259.
4. R. Borghi, B. Guidi, L. Lunazzi, G. Placucci, *J. Org. Chem.*, 1996, **61**, 5667.
5. R. Borghi, L. Lunazzi, G. Placucci, P. J. Krusic, D. A. Dixon, N. Matsuzawa, M. Ata, *J. Amer. Chem. Soc.*, 1996, **118**, 7608.
6. R. Borghi, L. Lunazzi, G. Placucci, P. J. Krusic, D. A. Dixon, L. B. Knight, *J. Phys. Chem.*, 1994, **98**, 5395.
7. O. G. Kalina, B. L. Tumanskii, V. V. Bashilov, A. L. Chistyakov, I. V. Stankevich, V. I. Sokolov, T. J. S. Dennis and R. Taylor, *J. Chem. Soc., Perkin Trans. 2,* 1999, 2655.
8. R. Borghi, B. Guidi, L. Lunazzi, G. Placucci, *J. Org. Chem.*, 1996, **61**, 5667.
9. B. L. Tumanskii, V. V. Bashilov, N. N. Bubnov, S. P. Solodovnikov, V. I. Sokolov, *Russ. Chem. Bull.*, 1995, **44**, 1771.
10. B. L. Tumanskii, V. V. Bashilov, N. N. Bubnov, S. P. Solodovnikov, V. I. Sokolov, Bull. *Russ. Acad. Sci.*, 1992, **41**, 1521.
11. B. L. Tumanskii, O. G. Kalina, V. V. Bashilov, unpublished results.
12. B. L. Tumanskii, V. V. Bashilov, N. N. Bubnov, S. P. Solodovnikov, V. I. Sokolov, *Russ. Chem. Bull.*, 1993, **42**, 203.
13. R. G. Gasanov, V. V. Bashilov, O. G. Kalina, B. L. Tumanskii, *Russ. Chem. Bull.*, 2000, **49**, 1642.
14. J. R. Morton, F. Negri, K. F. Preston, *Chem. Phys. Lett.*, 1994, **218**, 467.
15. J. R. Morton, K. F. Preston, F. Negri, *Chem. Phys. Lett.*, 1994, **221**, 59.
16. O. V. Boltalina, A. Ya. Borschevskii, L. N. Sidorov, J. M. Street, R. Taylor, *Chem. Commun.*, 1996, 529.
17. P. R. Birkett, A. G. Avent, A. D. Darwish, H. W. Kroto, R. Taylor, D. R. M. Walton, *J. Chem. Soc., Chem. Commun.*, 1995, 683.

Chapter 4

Radical Addition to C_{76}

Analysis of the publications concerned with fullerene chemistry shows that most studies are devoted to C_{60} and C_{70} fullerenes. Higher fullerenes have been little studied yet because of the difficulties associated with their synthesis and isolation from the fullerene mixtures. Those microscopic quantities that can be isolated and reliably characterized are inadequate to carry out target-directed syntheses.[1]

Moreover, C_{70} comprises 5 types of nonequivalent carbon atoms with different reactivity.[2] There are 19 types of nonequivalent carbon atoms in fullerene C_{76}, which seem to have different reactivity toward radicals.[3] A superposition of signals of all isomers resulting from addition of carbon-centered radicals to C_{76} can be expected in EPR spectrum.

In this situation, it is extremely difficult to identify characteristic features of the electronic structure and the chemical behaviour of these compounds. For this purpose, one can use the "phosphoryl EPR spectroscopy" procedure elaborated for C_{60} and C_{70}, according to which a phosphoryl group having attached through a radical process serves as a "paramagnetic reporter".[4]

Analysis of the spectra of the spin adducts formed by higher fullerenes with a phosphoryl group and comparison of the results obtained experimentally with the results of quantum-chemical calculations permit one to draw conclusions about the most reactive sites in the fullerene cage and the pattern of spin electron density distribution.

One of C_{76} isomers is the first chiral object in the fullerene series; in addition, according to calculations, it is also the most stable among the possible isomers.[5] The cage of this isomer consists of fragments resembling C_{60} in structure and planar sections representing coupled hexagons (Fig. 3, Chapter 1). It is this isomer that was used in the experiments described below.

It should be noted that radical reactions of C_{76} fullerene have not been studied. The few chemical transformations known for this compound are osmylation,[6] cycloaddition,[7] hydrogenation,[8] and the formation of methylene derivatives.[9] Only the last-mentioned products were characterized.

Both [70]- and [76]fullerenes have some similar features, namely an elongated shape, highly curved polar regions and a more planar and aromatic equatorial region, which has varying curvature in two orthogonal planes parallel to the long axis in the case of [76]fullerene.

[76]-Fullerene has four bonds of high π-bond order in the polar region (the 1,6-, 2,3-, 4,5-, and 14,15-bonds) and this leads to the expectation of the ready formation of seven radical adducts (the 1- and 6-positions are equivalent). In these polar regions, the spin densities adjacent to the addend will be high and hence the probability of dimerisation should be low.[10]

In order to ensure the predominant formation of monoadducts, the reaction was carried out in the presence of $(CF_3)_3COH$. This resulted in the appearance of seven groups of EPR signals (a_P = 40.2 G, g = 2.0032; a_P = 42.25 G, g = 2.0028; a_P = 44.5 G, g = 2.0028; a_P = 45.75 G, g = 2.0025; a_P = 52.5 G, g = 2.0029; a_P = 55.0 G, g = 2.0017) (Fig. 1).

When irradiation has been switched off, signals due to six isomers remain in the spectrum, whereas the seventh one disappears upon dimerization.[10]

To aid location of the sites of the addition of the phosphoryl radicals, calculations of ΔH^o_f for $^{\bullet}C_{76}P(O)(OPr^i)_2$ species were carried out using the MNDO/PM3 method at the ROHF level (Table 1). Spin-density distributions for all possible 19 isomeric spin-adducts were calculated by B3LYP/DFT method with the geometries optimized by PM3 (Table 2).[10,11]

The sums of the valence angles at each non-equivalent carbon of [76]fullerene were also optimized in order to determine the "convexity degree" of the carbon center in the fullerene framework.

The greater the difference between angle sum and 360°, the more convex is the framework in the vicinity of this carbon, and the greater the possibility of the radical addition to the local double bond since the latter will be less delocalized (higher π-bond order). Moreover, less distortion of the framework will be produced on conversion of the carbon hybridization from sp^2 to sp^3 upon addition.[12]

As noted above, the four bonds in [76]fullerene having the highest (and similar) calculated π-bond orders are the 14,15-, 4,5-, 1,6- and 2,3-bonds. There is an appreciable gap in bond order to the next group *viz.*, the 29,30-, 11,28- and 12,13-bonds (which as a group also have similar bond orders) and then further gaps to other groups of bonds. Since the 1- and 6-positions are identical, then seven radical adducts could be expected on the basis of kinetic stability (the number observed) with addition to carbons C-14, C-15, C-4, C-5, C-l, C-2, C-3.

$$Hg[P(O)(OR)_2]_2 \xrightarrow{h\nu} 2\,{}^{\bullet}P(O)(OR)_2 + Hg$$

$$C_{76} + {}^{\bullet}P(O)(OR)_2 \rightarrow {}^{\bullet}C_{76}P(O)(OR)_2$$

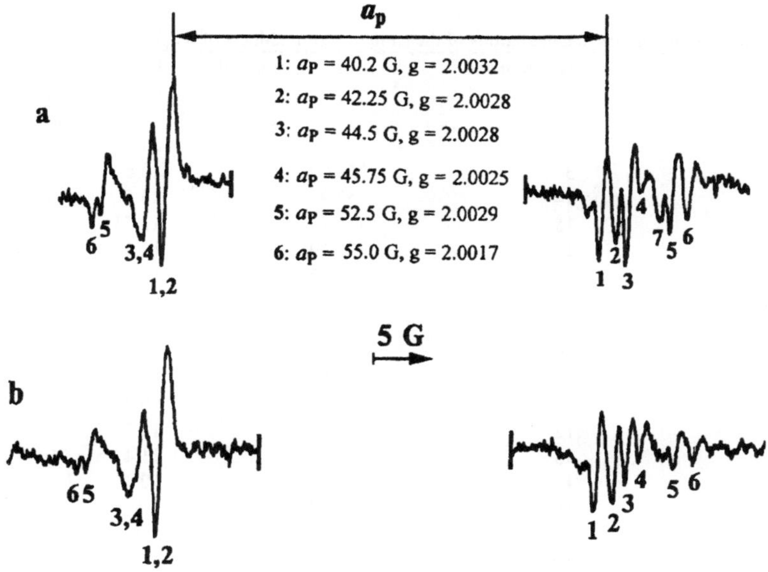

Figure 1. EPR spectrum of the spin-adducts of reaction of phosphoryl radicals with C_{76} at 300 K. (a) Under UV irradiation; (b) without UV-irradiation.

The calculated thermodynamic stabilities of all possible isomers (Table 1) also indicate that phosphoryl should attach itself to the following atoms (in the order of increasing ΔH°_f): C-3, C-14, C-15, C-1, C-2, C-4, C-5; *i.e.*, C > I > J > A > B > D > E (Fig. 2, 3).

The fragments of the framework containing these atoms also comprise seven of the nine atoms having the greatest convexity.

Addition to C-11, C-16, C-32, C-30, C-17, C-31, C-12, C-13 is also thermodynamically reasonably favourable but these sites are calculated to be more planar, thereby lowering the probability of addition (Fig. 4). The remaining sites are predicted to have either low stability or low convexity, or both.

The question then arises as to why one of the radicals is less stable than the others. Since, there is no conflict between the kinetic and thermodynamic orders, in contrast to the [70]fullerene, it is reasonable to suppose that either:

(a) only these seven sites are involved, but one of them, due to greater curvature of the fullerene surface, is able to dimerize like [70]fullerene, or (b) one of the favorable radicals is not formed and addition takes place elsewhere in the molecule where there is greater delocalization of the unpaired electron.

Table 1. Semi-empirical calculations of the isomeric spin-adducts of $^{\bullet}C_{76}P(O)(OPr^i)_2$.

Isomer	Addition site	$\Delta(\Delta H^{\circ}_f)$, kcal mol^{-1}	Convexity (360^0 minus sum of the carbon valency angles)
C	3	752.60	13.6
I	14	752.96	13.5
J	15	753.10	13.25
A	1	755.19	13.5
B	2	755.92	13.5
D	4	756.18	11.4
E	5	758.02	11.3
P	30	759.47	8.7
Q	31	759.55	7.9
R	32	759.66	10.4
G	12	760.03	11.7
H	13	760.75	11.4
L	17	761.23	10.4
K	16	761.37	8.5
F	11	762.28	8.7
N	19	769.11	5.0
O	29	770.31	4.1
S	33	770.75	6.9
M	18	773.22	5.1

There should be strong *para* delocalization of the spin density to C-36 for addition to C-16, and to C-35 for addition to C-17, and this is confirmed by the calculations in Table 2.

Addition/delocalization involving C-18/C-34 will, by contrast, be strongly disfavored by the disadvantageous convexity, *i.e.*, the strain that will result on forming an sp^3 carbon at the junction of three hexagons is confirmed by the low calculated stability (Table 1). Both the addition and delocalization site involve a three-hexagon junction.

Four different carbons are located at three-hexagon junctions, *viz.*, C-18, C-19, C-29, and C-33.

Delocalization of an unpaired electron to adjacent addition sites will be also unfavourable, and this accounts for the δ delocalization found for addition to C-ll, C-16, C-17, C-30, and C-31.

Overall, the most probable candidate is isomer P taking into account all the stability and convexity factors and the calculated spin density at the δ carbon, if it is the unstable isomer that involves δ delocalization.

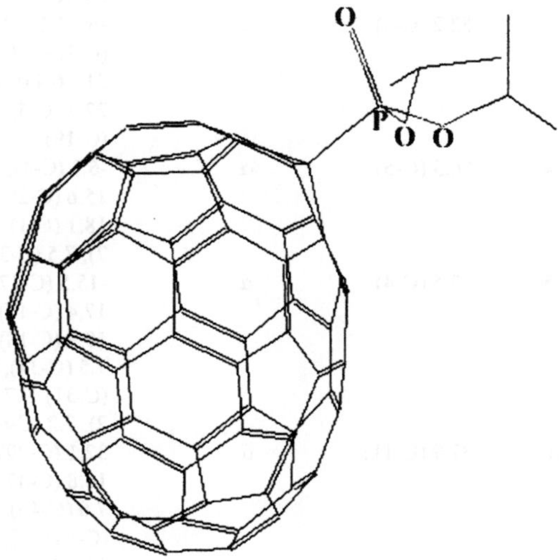

Figure 2. Hypotetical structure for the most stable (C) spin-adduct of C_{76} with phosphoryl radical.

Table 2. Calculated (B3LYP) spin density distributions for the 19 isomers of $^{\bullet}C_{76}P(O)(OPr^{i})_2$.

Isomer	Addition site	Principal spin density and location	Type	Spin densities (>6%) at other sites
C	3	54.9 (C-2)	α	-13.9 (C-1), 19.8 (C-6), 12.2 (C-8), 7.4 (C-4), -13.3 (C-12), 9.7 (C-11), 14.4 (C-13), 11.3 (C-29)
I	14	53.6 (C-15)	α	-12.9 (C-3), 21 (C-2), 9.2 (C-6), -15.5 (C-16), 11.7 (C-34), 15.5 (C-17), -7.2 (C-18), -7.1 (C-35), 16.5 (C-36)
J	15	55.6 (C-14)	α	-15.7 (C-32), 10.3 (C-33), -6.6 (C-52), 15.6 (C-51), -7.5 (C-50), 7.4 (C-67), 18.6 (C-31), -10.9 (C-13), 14.9 (C-12), 7 (C-28)
A	1	56.4 (C-6)	α	20.9 (C-4), -15.3 (C-5), 15.7 (C-7), 8.5 (C-10), -6.2

Isomer	Addition site	Principal spin density and location	Type	Spin densities (>6%) at other sites
				(C-1), 6.3 (C-2), 22.5 (C-8), -6.9 (C-24), 12.2 (C-23), 6.6 (C-18)
B	2	52.2 (C-3)	α	-6 (C-2), -14.9 (C-4), -14.9 (C-15), 23.6 (C-14), 6.4 (C-31), 6.4 (C-51), 6.1 (C-33), 22.7 (C-5), -6 (C-20), 6.9 (C-19)
D	4	56.3 (C-5)	α	-6.3 (C-4), 16.5 (C-19), -15.6 (C-20), 15.8 (C-40), 18.1 (C-1), -14 (C-6), 6 (C-7), 7.5 (C-3), 9.1 (C-21)
E	5	52.5 (C-4)	α	-15.5 (C-17), -13.1 (C-3), 17.4 (C-16), 16.2 (C-35), 10.9 (C-18), -7.2 (C-36), 7.5 (C-34), 7.6 (C-53), 6.1 (C-31), 9.7 (C-13), 20.3 (C-2), 7.2 (C-6)
P	30	37.9 (C-11)	δ	32.1 (C-29), -11.1 (C-48), 16.8 (C-47), -15.7 (C-28), -7.6 (C-46), 11.1 (C-45), -6 (C-44), 12(C-25), -8 (C-26), 15.4 (C-9), 10.9 (C-10), -9.5 (C-12), 14.4 (C-13), 6.4 (C-15)
Q	31	37.1 (C-52)	δ	30.8 (C-32), 14.4 (C-50), 17.2 (C-54), -10.9 (C-53), -14.5 (C-33), 9.9 (C-51), 7.4 (C-66), -10.1 (C-14), 15.4 (C-15), -7.3 (C-16), 10.2 (C-17), -6 (C-18), 12.6 (C-36), -8.2 (C-35)
R	32	38.8 (C-31)	α	-14.8 (C-50), 28.2 (C-51), 15.5 (C-67), -9.5 (C-68), 6.6 (C-65), 14 (C-47), 7.3 (C-45), 7.2 (C-25), -7.2 (C-28, -8.4 (C-48), 14.6 (C-29), 14.5 (C-11), 6.2 (C-49), -8 (C-66), -12.7 (C-30), -7.5 (C-52), 12.1 (C-33), 7 (C-13)
G	12	42.7 (C-13)	α	-12.9 (C-14), 21.3 (C-15), -15.3 (C-30), 23.1 (C-29), -7.3 (C-48), 9.5 (C-47), -8.5 (C-28), 7.2 (C-45), 7.3 (C-9), 14.7 (C-11), 8.4 (C-2), -6 (C-16), 7.3 (C-17), 6.1

Isomer	Addition site	Principal spin density and location	Type	Spin densities (>6%) at other sites
				(C-36)
H	13	44.4 (C-12)	α	17.3 (C-30), -9.6 (C-29), 21.7 (C-28), 8.9 (C-46), -6 (C-27), -14.5 (C-1), -12.1 (C-2), 15.9 (C-3), -6 (C-4), 13.6 (C-5), 6 (C-52)
L	17	34.3 (C-35)	δ	15 (C-14), -9.8 (C-15), 30.6 (C-16), 16.9 (C-18), 6.2 (C-20), 12.1 (C-31), -7.9 (C-32), -14 (C-34), -10.2 (C-36), 11.6 (C-51), -8.4 (C-52), 17.9 (C-53), -11.5 (C-54), 6.9 (C-70)
K	16	37.8 (C-36)	δ	26.6 (C-17), -14.2 (C-5), 12.8 (C-40), 12.8 (C-38), -12.1 (C-37), 13.2 (C-57), -9.6 (C-35), 15.5 (C-34), -13.7 (C-18), 6 (C-19), -7 (C-20), 6.8 (C-42), 6.3 (C-61), -7.6 (C-39), 8.2 (C-55), -8.8 (C-4)-6.4 (C-58)
F	11	41.7 (C-30)	δ	30.5 (C-28), -11.9 (C-31), 7.1 (C-50), 6.5 (C-66), -15.5 (C-29), 9.8 (C-46), 8.8 (C-27), 8.2 (C-26), 13 (C-12), -8.9 (C-13), 6.8 (C-3), 15.2 (C-32), -6.5 (C-33), 13.2 (C-52)
N	19	55.6 (C-20)	α	8.5 (C-2), 19.2 (C-4), -14.2 (C-5), 6.2 (C-7), 6.6 (C-18), -7.1 (C-19), -15 (C-21), 20.6 (C-22), -6.4 (C-23), 11.3 (C-24), 9.8 (C-39), -6.3 (C-40),
O	29	53.6 (C-30)	α	18.5 (C-12), 10.6 (C-50), -6.3 (C-51), -14.7 (C-31), -13.3 (C-13), 8.8 (C-28), 10.8 (C-3), -6.6 (C-33), 14.4 (C-52), 6.5 (C-54), 15.6 (C-32)
S	33	48.3 (C-52)	α	-16.3 (C-51), 21.9 (C-50), 11.2 (C-66), -6.9 (C-49), -12.1 (C-31), 22.7 (C-32), -6 (C-14), 7.3 (C-15), 10.1 (C-34), 19.3 (C-54), -12.3 (C-53), 6 (C-56)
M	18	44.8 (C-36)	α	-6.5 (C-4), 8.1 (C-5), -13

Isomer	Addition site	Principal spin density and location	Type	Spin densities (>6%) at other sites
				(C-16), 25.8 (C-17), -7.1 (C-18), 7.8 (C-19), 7.3 (C-32), -6 (C-33), 23 (C-34), -15.7 (C-35), -11.8 (C-37), 15.1 (C-38), 7.2 (C-52), 10.3 (C-57)

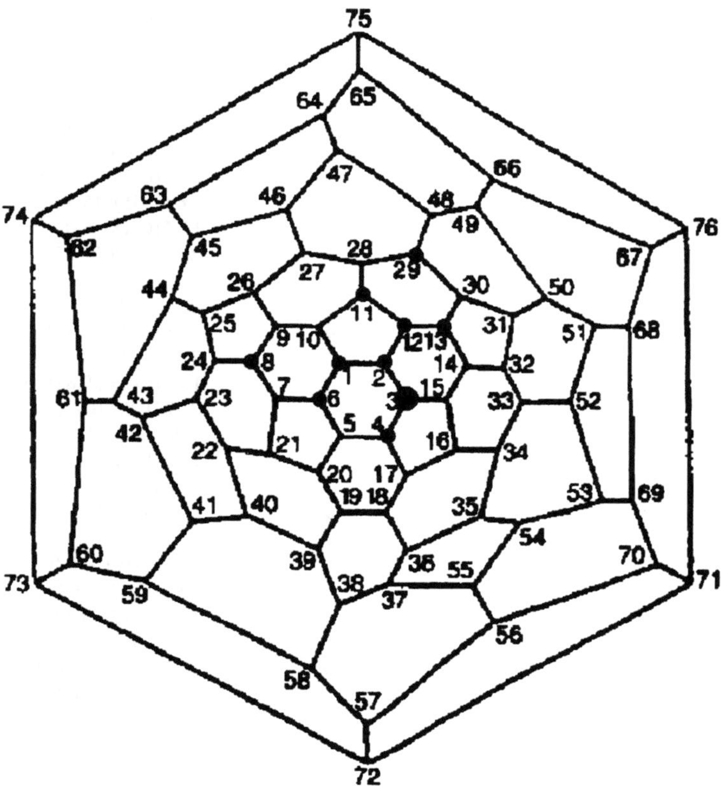

Figure 3. Spin density distribution for the spin-adduct of C_{76} and phosphoryl radical connected to C-3 atom.

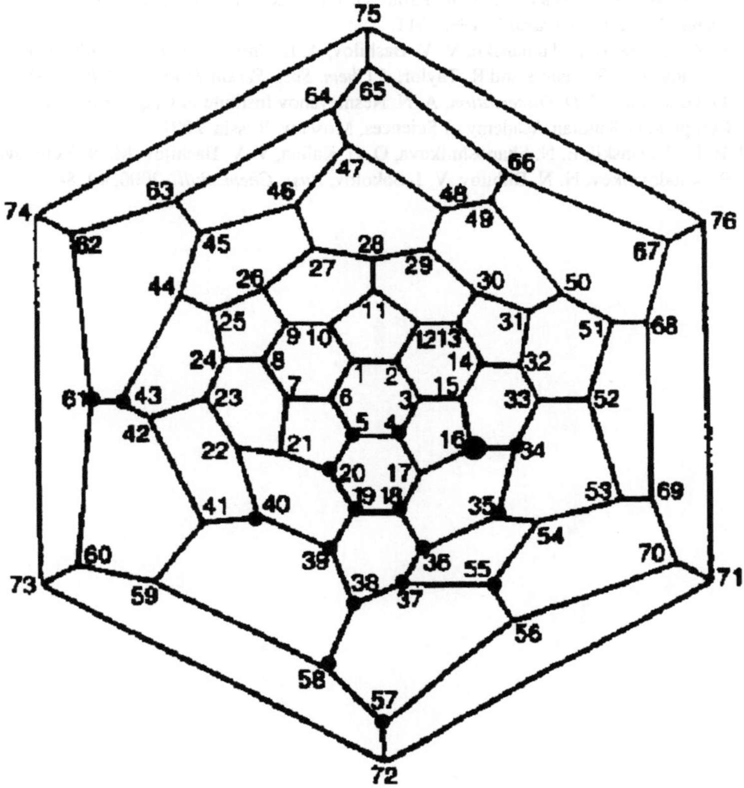

Figure 4. Spin density distribution for the spin-adduct of C_{76} and phosphoryl radical connected to C-16 atom.

References

1. R. Taylor, *Lecture Notes on Fullerene Chemistry: A Handbook for Chemists*, Imperial College Press, London, 1999.
2. R. Borghi, L. Lunazzi, G. Placucci, P. J. Krusic, D. A. Dixon, N. Matsuzawa, M. Ata, *J. Amer. Chem. Soc.*, 1996, **118**, 7608.
3. R. Taylor, *J. Chem. Soc., Perkin Trans. 2*, 1993, 813.
4. B. L. Tumanskii, V. V. Bashilov, O. G. Kalina, *Chemistry Reviews*, 2001, in press.
5. C. Thilgen, A. Herrmann, F. Diederich, *Helv. Chim. Acta*, 1997, **80**, 183.
6. J. M. Hawkins, A. Meyer, *Science*, 1993, **260**, 1918.
7. A. Herrmann, F. Diederich, *Helv. Chim. Acta*, 1996, **79**, 1741.
8. A. D. Darwish, H. W. Kroto, R. Taylor, D. R. M. Walton, *J. Chem. Soc., Perkin Trans. 2*, 1996, 415.

9. P. R. Birkett, A. D. Darwish, H. W. Kroto, G. J. Langley, R. Taylor, D. R. M. Walton, *J. Chem. Soc., Perkin Trans. 2*, 1995, 511.
10. O. G. Kalina, B. L. Tumanskii, V. V. Bashilov, A. L. Chistyakov, I. V. Stankevich, V. I. Sokolov, T. J. S. Dennis and R. Taylor, *J. Chem. Soc., Perkin Trans. 2,* 1999, 2655.
11. O. G. Kalina, *Ph.D. Dissertation*, A. N. Nesmeyanov Institute of Organoelement Compounds, Russian Academy of Sciences, Moscow, Russia, 2001.
12. B. L. Tumanskii, E. N. Shaposhnikova, O. G. Kalina, V. V. Bashilov, M. N. Nefedova, S. P. Solodovnikov, N. N. Bubnov, V. I. Sokolov, *Russ. Chem. Bull.*, 2000, **49**, 843.

Chapter 5

Radical Addition to Fullerene Derivatives

The addition of metal complexes, methylene groups, or free radicals to fullerenes results in deformation of the fullerene cage, considerable decrease in its symmetry, and the appearance of several groups of non-equivalent carbon atoms. This brings about the question of the degree to which functional groups of various chemical natures influence the deformation of the fullerene core, the regiochemistry for addition of other radicals, and the type of binding and mutual arrangement of the groups attached to fullerene.[1]

This question can be answered resorting to joint analysis of the EPR spectra of the spin adducts of radicals with these derivatives and the results of theoretical calculations of the resulting compounds.

1.1 Radical Reactions with Metal Complexes of Fullerenes

The initial organometallic fullerene derivatives used in these studies were L_2MC_{60}, where M = Pt, Pd;[2] L = PPh_3; $C_{60}IrH(CO)(PPh_3)_2$,[3] and $C_{70}Pd(DIOP)$.[4]

UV irradiation of a saturated toluene solution of L_2MC_{60} in the presence of $Hg[P(O)(OPr^i)_2]_2$ at room temperature in the resonator of an EPR spectrometer results in superposition of the spectra of at least five radicals differing in the HFC constants of the unpaired electron with the ^{31}P nuclei (70—55 G) and in *g*-factors (2.001—2.003).[5] The parameters of the predominant signals fully coincide with those for radical $^{\bullet}C_{60}P(O)(OPr^i)_2$. After 3—5 minutes of irradiation, only the spectrum of this radical is retained. Evidently, the phosphoryl radical attacks the L_2MC_{60} complex non-selectively.

At the beginning of the reaction, when the concentration of the metal complex is high and no C_{60} is present in the solution, the phosphoryl radicals undergo addition to the fullerene metal complex at different distances from the metal, giving rise to spin adducts with different *g*-factors and *hfc* constants. However, the attack of the phosphoryl radical not only on the metal but on the double bond closest to the M–C bonds yields a radical in which the M–C bonds are weakened; the complex eliminates L_2M to give radical $^{\bullet}C_{60}P(O)(OPr^i)_2$.[5]

Metal-containing phosphorylfullerenyl radicals have been prepared by a independent synthesis, *i.e.* upon the addition of PtL_2 to the dimers of phosphorylfullerenyl radicals. The complex $Pt(PPh_3)_4$ was added under argon to a toluene solution containing an equimolar amount of the phosphorylfullerenyl radical dimer:

$$Hg[P(O)(OPr^i)_2]_2 \xrightarrow{h\nu} 2\,{}^{\cdot}P(O)(OPr^i)_2 + Hg$$

$$C_{60} + {}^{\cdot}P(O)(OPr^i)_2 \longrightarrow {}^{\cdot}C_{60}P(O)(OPr^i)_2$$

$$2\,{}^{\cdot}C_{60}P(O)(OPr^i)_2 \longrightarrow (OPr^i)_2(O)PC_{60}\text{–}C_{60}P(O)(OPr^i)_2$$

$$RC_{60}\text{–}C_{60}R + 2\,Pt(PPh_3)_4 \longrightarrow (R(PPh_3)_2PtC_{60}\text{–}C_{60}Pt(PPh_3)_2R$$

$$R = P(O)(OPr^i)_2$$

$$(R(PPh_3)_2PtC_{60}\text{–}C_{60}Pt(PPh_3)_2R \xrightarrow{h\nu} 2\,{}^{\cdot}C_{60}Pt(PPh_3)_2R$$

After the reaction, the sample was exposed to visible light (620–680 nm), to avoid generation of additional phosphoryl radicals and to affect only the dimer of fullerenyl radicals, in which the energy of the C–C bond is relatively low (see Chapter 2).

Exposure of the sample to visible light induces dissociation of the dimer, giving rise to metal-containing fullerenyl radicals (a_P = 58.5 G, g = 2.0017; a_P = 64.7 G, g = 2.0025; a_P = 65.7 G, g = 2.0028, and a_P = 68.0 G, g = 2.0032) (Fig. 1).

In the case of the iridium complexes, $(\eta^2\text{-}C_{60})[IrH(CO)(PPh_3)_2]$, where Ir is linked more firmly to fullerene, metal-containing radicals (a_P = 66.7 G, g = 2.0026; a_P = 64.25 G; g = 2.0037; a_P = 59.25 G; g = 2.0027) can be detected upon direct reaction of phosphoryl radicals with the metal complexes.[6]

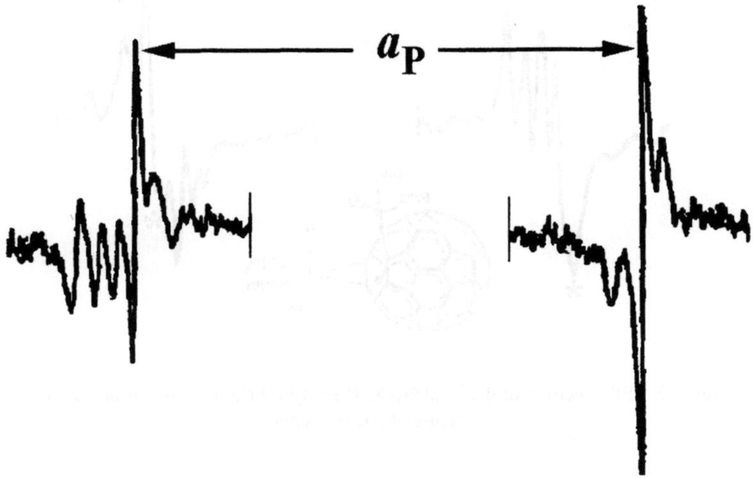

Figure 1. EPR spectrum of isomers of the radical $^{\bullet}C_{60}Pt(PPh_3)_2P(O)(OPr^i)_2$ at 293 K.[5]

The width of the spectral line of one of the isomer (*, Fig. 2) formed ($\Delta H_{p\text{-}p}$ = 2.5 G) is 2.5 times greater than those for other isomers, the intensity of the EPR spectrum of this isomer being greater than the total intensity of the spectra of the other metal-containing phosphorylfullerenyl radicals.

The increase in the width of the line corresponding to this isomer is due to the *hfc* of the unpaired electron with the iridium, triphenylphosphine phosphorus, and hydrogen nuclei.

In other words, the phosphoryl radical has attached to a position close to metal (spaced by 2 or 3 three bonds from it).

The mild conditions under which 3,5-dimethylphenylmethyl radicals undergo addition to C_{60} in the presence of the dimer $[F(C_6H_4)C(CF_3)_2]_2$ (see Chapters 1, 2) make it possible to study the radical polyaddition of carbon-centered radicals to $(\eta^2\text{-}C_{60})[IrH(CO)(PPh_3)_2]$ and to estimate the relative rate of the addition to it compared to C_{60}.[7]

It was found that, when the reaction is carried out under the same conditions as that for C_{60}, the signal intensity in the EPR spectrum of the resulting spin adduct is about five times lower than that for the spin adduct formed by 3,5-dimethylphenylmethyl radicals and C_{60}.

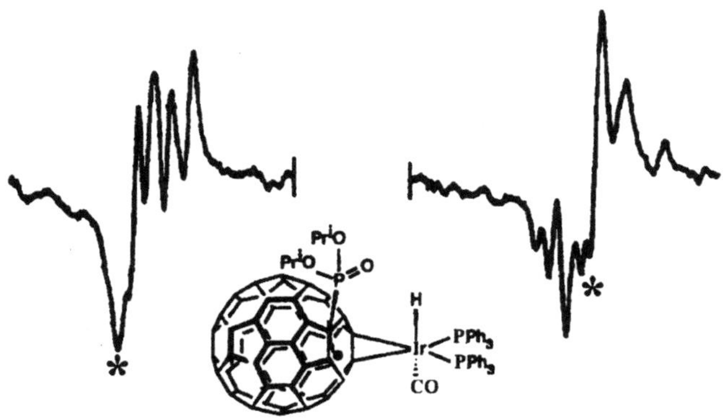

Figure 2. EPR spectrum of the $^{\bullet}C_{60}IrH(CO)(PPh_3)_2P(O)(OPr^i)_2$ isomers and a possible structure of one of them (*).[6]

Only when the metal complex concentration in a mesitylene solution is five times as high as the C_{60} concentration, a similar pattern of adduct accumulation is observed in the spectrum. This implies that the addition to fullerene of the electron-donating organometallic group, $IrH(CO)(PPh_3)_2$, decreases the rate of the reaction of 3,5-dimethylphenylmethyl radicals with the substituted fullerene. It should be noted that the *g*-factor ($g = 2.0028$) of the metal-containing fullerenyl radical differs from the *g*-factor of the spin adducts of 3,5-dimethylphenylmethyl radical with C_{60} ($g = 2.0021$).

The reason for this difference is that 3,5-dimethylphenylmethyl radicals attach themselves to the same hemisphere to which the metal-containing group is linked and the unpaired electron interacts with the metal:[7]

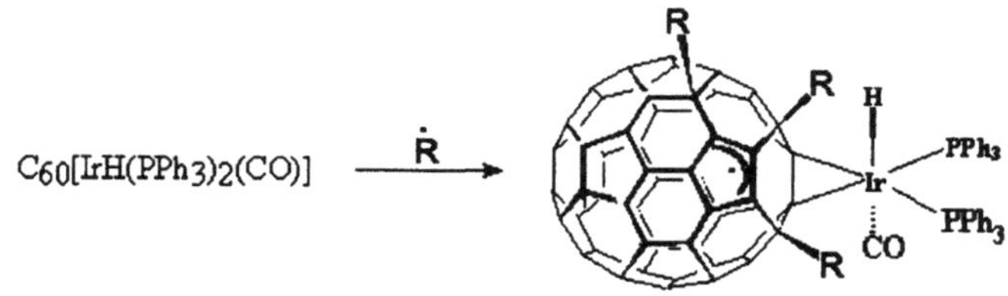

Indeed, heating the metal-containing fullerenyl radical up to the temperature of decomposition of the metal complex (370 K) over a period of 6 min results in a shift of the EPR line toward the *g*-factor corresponding to the spin-adduct formed from C_{60} (Fig. 3).

Photochemical phosphorylation of $C_{70}Pd(DIOP)$ in the resonator of an EPR spectrometer gives rise to an unusual spectrum (Fig. 4).[4] In addition to the three doublets marked by asterisks and corresponding to regioisomers of the spin adducts of phosphoryl radicals with demetallated C_{70}, the spectrum exhibits three types of doublets of doublets (A–C) with the following magnetic-resonance parameters ($a(^{31}P_1)$ = 30.0 G, $a(^{31}P_2)$ = 54.0 G, g = 2.0112; $a(^{31}P_1)$ = 30.0 G, $a(^{31}P_2)$ = 53.3 G, g = 2.0040; $a(^{31}P_1)$ = 30.0 G, $a(^{31}P_2)$ = 53.5 G, g = 2.0043). The fact of coupling of the unpaired electron with two phosphorus nuclei, together with the fact that the *g*-factor differs appreciably from those of the isomers of phosphorylfullerenyl radicals, indicate that the given spectrum corresponds to the adduct of metal-centered radicals formed upon rupture of the C–M bond upon the attack of the phosphoryl radical on the metal.

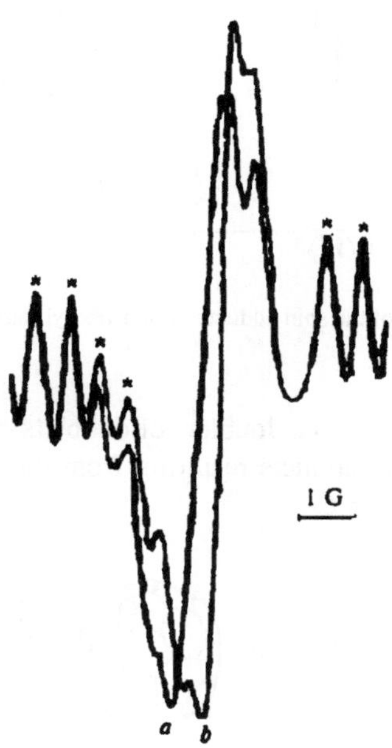

Figure 3. EPR spectra of the metal-containing fullerenyl radical in mesitylene at 300 K (a) and demetallated fullerenyl radical during heating at 370 K for 6 min (b). The components of the EPR spectrum of radical $F(C_6H_4)(CF_3)_2C^{\bullet}$ are marked by asterisks.[7]

The high value of the *hfc* constant with the phosphorus nucleus corresponds to the phosphoryl group, while the smaller constant refers to one of the phosphorus nuclei in the ligand environment and is virtually equal to the *hfc* constant with the phosphorus nucleus of one triphenylphosphine group in the spin adduct of the platinum-centered radical with C_{60}.

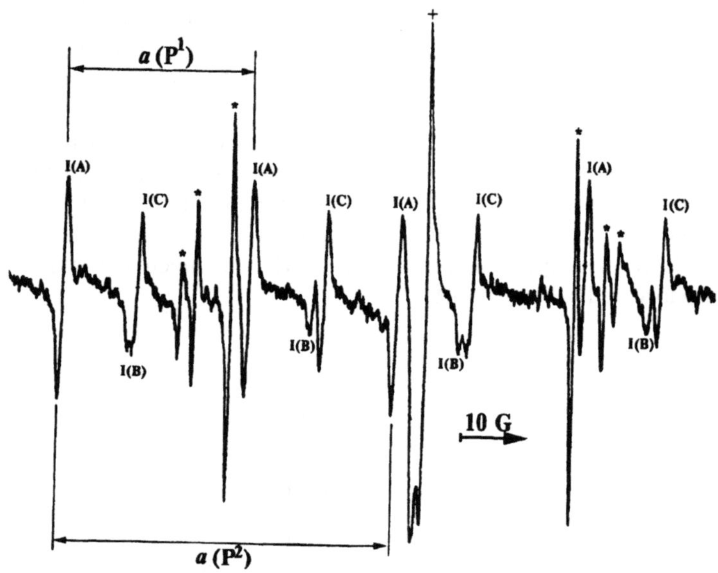

Figure 4. EPR spectrum of the spin adducts of phosphoryl radicals with $C_{70}Pd(DIOP)$.[4]

The fact that there are three doublets of doublets of this type is due to the presence of isomers, *i.e.* adducts resulting from the addition of Pd to non-equivalent carbon atoms in C_{70}:

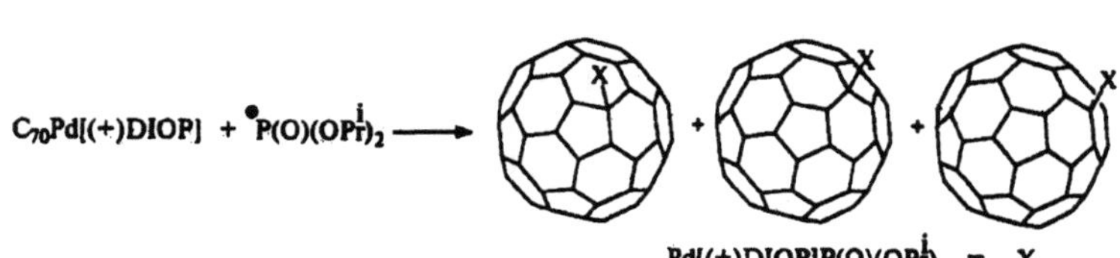

1.2 Reactions of Radicals with Organic Derivatives of C_{60}

Unlike the metal complexes of fullerenes, in which the metal–fullerene bond is relatively labile, methano derivatives of fullerene are much more stable and do not degrade under the conditions of photochemical generation of radicals.

1.2.1 Addition of Hydroxyl Radicals

The reaction of $^{\bullet}OH$ with water-soluble $C_{60}[C(COOH)_2]_n$, n = 2-6, has been studied.[8] An excess of the reactant was used to produce monoadducts of $OH^{\bullet}$ to compounds studied.

Transient electronic spectra of monohydroxylated fullerenyl radicals, derived from these compounds reacting with hydroxyl radicals at a diffusion-controlled rate, showed a unimolecular growth in absorption with no significant change in spectral profile. It was suggested that the reason for this growth is migration of the hydroxyl group to a more stable position.[8]

To determine the regiochemistry of hydroxyl radical addition to derivatives under study, semiempirical calculations of formation enthalpies and indices of free valence were performed.

A hydroxyl radical was added to each of the symmetry-distinct vertices of the anion to form the closed-shell species, whose structure was then fully geometrically optimised to obtain its energy. This approach assumes that the relative thermodynamic stabilities of these closed-shell hydroxide anions parallel the stabilities of the structurally related hydroxyfullerenyl radicals.[8]

The free valence index (**F**$_r$) measures the ease of the attack of free radicals at an atom in a conjugated π-system. It was shown that the final product arising from the radical addition could be predicted by assuming initial attack at the position of highest free valences.

Semiempirical calculations on $C_{60}[C(COOH)_2]_3OH^-$ anion, derived from the reaction of $C_{60}[C(COOH)_2]_3^{-\bullet}$ with hydroxyl radical, demonstrate a relation between free valence index for the radical attack at a particular position and the stability of the corresponding product (Table 1, Fig. 2 in Chapter 1).

Table 1. Data of semiempirical calculations on $C_{60}[C(COOH)_2]_3OH^-$ (e,e,e) isomer.[8]

Site of addition	Energy, kJ mol^{-1}	F_r index
10	17.8	0.1536
13	14.4	0.1461
7	33.4	0.1192
8	26.2	0.1295
28	17.7	0.1703

Site of addition	Energy, kJ mol^{-1}	F_r index
26	20.1	0.1734
30	10.7	0.1787
24	5.1	0.1768
29	17.7	0.1707
25	25.3	0.1369
27	5.6	0.1788
45	5.8	0.1790
47	7.5	0.1790
12	0.0	0.1739
46	11.1	0.1757
6	9.2	0.1721
9	12.1	0.1693
11	14.2	0.1626

The correlation for each isomer between the value of the free valence index and the heat of formation of the corresponding anion is reasonable and high values for the free valence indices correlate with low heats of formation of the product.

The ease of the attack of hydroxyl radical on a particular position is reflected in a greater calculated stability for the corresponding hydroxide anion.

For a highly reactive $^{\bullet}OH$, it is likely that the final energy of the product will correlate with the reactivity of the site of attachment.[8]

With such a small range of energies, the final radical product is a mixture of isomers. The pattern of stability of these species is that the most reactive positions toward an attacking radical are those where a hydroxy group has bound to a carbon adjacent to malonic acid group.[8]

They also suggest that after attack at the first encountered double bond, migration subsequently takes place to the position of the highest free valence.[8]

1.2.2 Addition of Phosphoryl Radicals

Upon radical phosphorylation of $C_{60}(CH_2)_2C_6H_2(OMe)_2$ where there are 16 groups of nonequivalent carbon atoms (2, 5, 7, 17; 3, 4, 18, 19; 8, 9, 15, 16; 11, 12, 22, 24; 13, 59, 21, 25; 10, 14, 20, 26; 27, 28, 56, 57; 23, 29; 52, 58; 39, 42, 50, 54; 38, 43, 49, 55; 40, 41, 51, 53; 31, 32, 47, 48; 36, 37, 44, 45; 30, 33, 35, 46; 34, 60), only 7 dominant signals in EPR spectrum were observed (1: a_P = 79.5 G; 2: a_P = 78.5 G; 3: a_P = 76.6 G; 4: a_P = 71.5 G; 5: a_P = 68.5 G; 6: a_P = 66.0 G; 7: a_P = 63.0 G) (Fig. 5,6).[9]

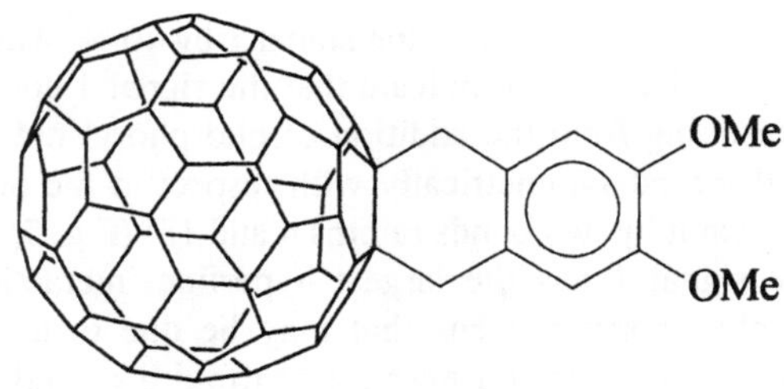

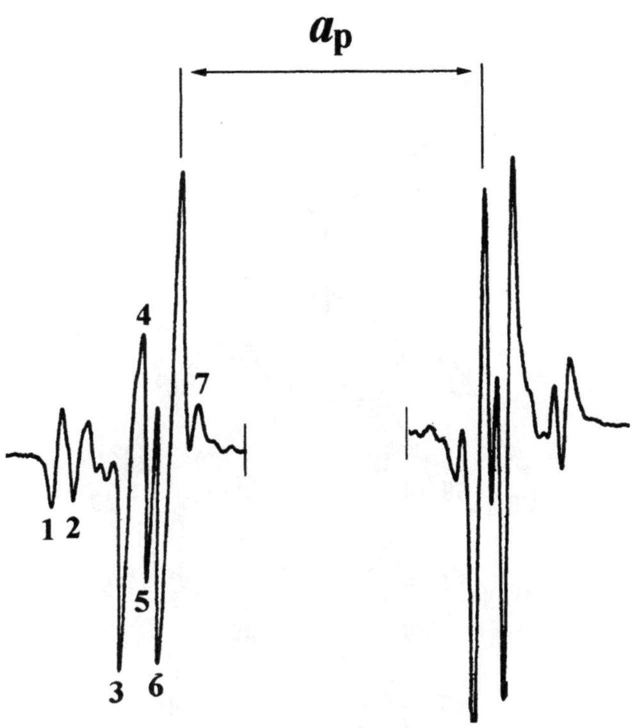

Figure 5. EPR spectrum of $^{\bullet}C_{60}(CH_2)_2C_6H_2(OMe)_2P(O)(OPr^i)_2$ isomers.[9]

When irradiation terminates, the signals 2–7 disappear, whereas the intensity of signal 1 decreases by 50%. The increase in temperature to 400 K leads to decrease in the intensity of the signal 1 by 40 %, while other signals remain unchanged. These data indicate that the signal 1 corresponds to two spin adducts resulting from the addition of phosphoryl radicals to scaffold carbon atoms disposed symmetrically with respect to the organic fragment and separated from it by two bonds (atoms 4 and 18) (Fig. 5, 6).

Indeed, the signal 1 has the largest hyperfine interaction of unpaired electron with phosphorus nucleus that may be due to a lesser extent of unpaired electron delocalization over the scaffold distorted in proximity to the organic fragment. The addition to atom 4, obviously, should result in less stable isomer than the addition to atom 18 that is manifested in its thermal dissociation upon heating and in the decrease of EPR signal.

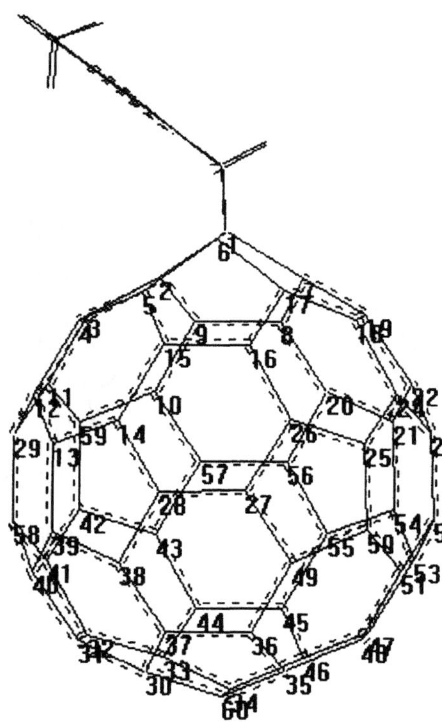

Figure 6. Structure of $C_{60}(CH_2)_2C_6H_2(OMe)_2$ with numbering of carbon atoms.

The stability of spin adducts corresponding to the signal 1 is associated with the shielding of carbon atoms carrying spin density by the bulky organic substituent and the attached phosphoryl fragment. This hinders the dimerization of resultant phosphorylfullerenyl radicals, which is known to be the main route of their destruction after irradiation ceased.

The isomers were further identified using EPR spectra of spin adducts recorded when temperature decreased from 300 to 200 K. A selective broadening of spectral lines of different isomers was observed due to hindering of diffusional rotation of molecule.

The rotation of the fullerenyl radicals around the symmetry Z-axis proves to be faster for the radicals where phosphoryl group is disposed at the opposite site of the fullerene scaffold with respect to the organic fragment (for example, atoms 30, 34, and 44):

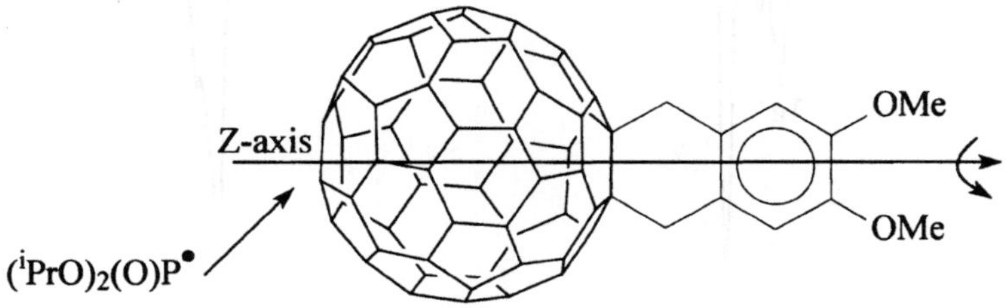

and their EPR spectra show conformational isomers, which differ in hyperfine interactions with P nucleus and in *g* factors, because of hindered rotation of the phosphoryl group around C–P bond (signals 3a, 3b; 6a, 6b) (Fig. 7).

Other lines in the EPR spectrum are broadened.

Quantum-chemical calculations of formation enthalpies were made for all isomeric spin adducts $^{\bullet}[C_{60}(CH_2)_2C_6H_2(OMe)_2]P(O)(OPr^i)_2$ by MNDO/PM3 method at ROHF level, and scaffold curvature was calculated at each of the nonequivalent carbons of $C_{60}(CH_2)_2C_6H_2(OMe)_2$ (Table 2).

The data obtained confirm the assignment of the signal 1 to the spin adducts resulting from attack of atoms 4 and 18 of fullerene scaffold, which have the maximal "curvature".

The addition to atoms 29, 34, 10, 43, 40, 44, 42, 30, 28, 31, 13, and 58 leads to the formation of thermodynamically more stable radicals. However, since EPR spectrum shows signals from 7 isomeric spin adducts, it is evident that the selective attack of carbon atoms with larger "curvature" occurs (Table 2).

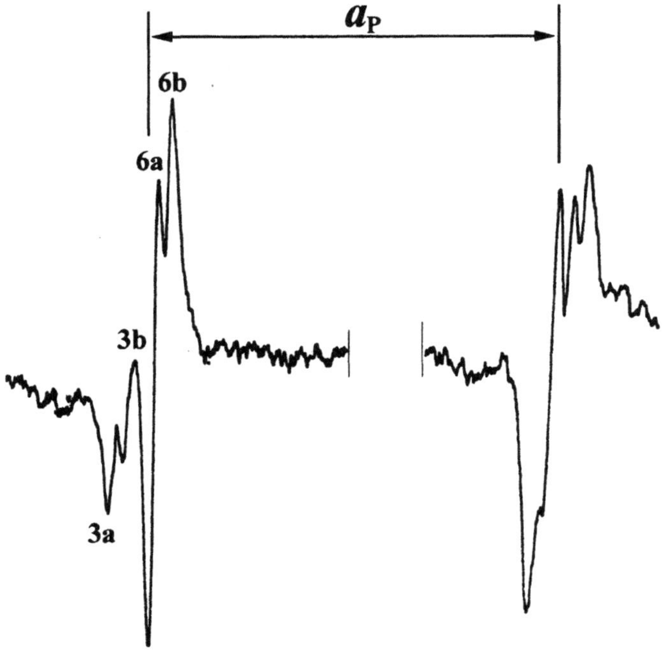

Figure 7. EPR spectrum of $^{\bullet}C_{60}(CH_2)_2C_6H_2(OMe)_2P(O)(OPr^i)_2$ isomers at 200 K.[9]

Table 2. Semi-empirical calculations of the isomeric spin-adducts of $^{\bullet}C_{60}(CH_2)_2C_6H_2(OMe)_2P(O)(OPr^i)_2$.

Site of addition	ΔH°_f, kcal mol^{-1}	Convexity (360^0 minus sum of the carbon valence angles)
29	551.45	11.533
34	551.62	12.012
10	551.75	13.527
43	552.46	12.313
40	552.49	12.07
44	552.54	11.943
42	552.91	12.097
30	553.1	12.051
28	553.15	12.043
31	553.35	12.025
13	553.73	12.012
58	553.9	11.639
18	554.89	15.339

Site of addition	$\Delta H°_f$, kcal mol^{-1}	Convexity (360^0 minus sum of the carbon valence angles)
12	556	11.427
4	562	15.339
9	562.34	8.851
2	566.09	6.766
17	566.09	6.766

For the structures optimized by PM3 method, spin density distribution and hyperfine interactions between unpaired electron and phosphorus nucleus were calculated by B3LYP/3-21 G method (Table 3).[9] Electron density maximum is located at the α carbon atom of the scaffold in all the isomers that is caused by the absence of flat sites in this molecule. However, the number of carbon atoms in the isomeric spin adducts participating in spin density delocalization is different (Fig. 8, 9). This is accounted for the distortion of fullerene polyhedron in the fullerene derivative as compared with the pristine C_{60}.

The isomers that differ in the position of attached phosphoryl radical have different hyperfine interactions with phosphorus nucleus, and the difference between the maximal and minimal interactions (82.3 – 66.5 G) is 15 G that virtually coincides with experimental data. This confirms the assumption that each signal in the EPR spectrum corresponds to the addition to the nonequivalent apex of the scaffold.

The replacement of methoxy group with hydroxy group in the attached organic fragment has no effect on the regiochemistry of radical addition and recorded EPR spectrum is similar to that described above.[9]

Table 3. Calculated (B3LYP) spin density distributions for the isomers of $^{\bullet}C_{60}(CH_2)_2C_6H_2(OMe)_2P(O)(OPr^i)_2$.[9]

Site of addition	Principal spin density and location	Spin densities (>6%) at other sites
10	50.1 (C-9)	-19.1 (C-2), 29.8 (C-3), -8 (C-4), 10.7 (C-5), 8.2 (C-7), -13.9 (C-8), -7 (C-11), 6.9 (C-14), -6.2 (C-19), 19.6 (C-20), -6.3 (C-21), 9.1 (C-22), 6 (C-52), 6.3 (C-57), 9.6 (C-59)
43	55.1 (C-44)	10.3 (C-30), -6 (C-31), 21.7 (C-32), -15.2 (C-33), 6.7 (C-42), -6.4 (C-43), -15.2 (C-45), 10.3 (C-53), -6 (C-54), 21.9 (C-55), 7.3 (C-57)

Site of addition	Principal spin density and location	Spin densities (>6%) at other sites
42	54.1 (C-41)	-6.2 (C-12), 9.6 (C-13), 6 (C-15), 22.3 (C-29), -14.9 (C-32), 22 (C-33), 6.6 (C-43), -15.1 (C-58), 8.5 (C-59), 9.3 (C-60)
60	56.5 (C-34)	8 (C-30), 21.2 (C-32), -15.2 (C-33), 7.8 (C-35), -6 (C-41), 9.7 (C-42), -15.1 (C-46), 21.2 (C-47), -6 (C-53), 9.8 (C-54), -6.8 (C-60)
13	50.7 (C-12)	-18.6 (C-4), 34.7 (C-5), -6.2 (C-13), 12.3 (C-14), -8.9 (C-15), -15.4 (C-29), 8.6 (C-39), -7.1 (C-41), 10.3 (C-42), 6.3 (C-44)
9 (15)	55.8 (C-14)	-6.8 (C-4), 11 (C-5), 22.3 (C-12), -15.3 (C-13), -7 (C-15), 6.8 (C-16), 21.7 (C-27), -15.2 (C-28), -6.7 (C-29), 9.7 (C-36), 10.3 (C-58),
40 (41)	54.5 (C-42)	15 (C-2), -7.6 (C-3), 18.8 (C-11), -15 (C-43), 23.9 (C-44), -6.1 (C-45), 10.8 (C-55), 6.8 (C-58), -14.6 (C-59),
30 (33)	55.0 (C-32)	7 (C-2), 9.5 (C-11), 23.1 (C-30), -15.3 (C-31), -6 (C-33), 7.4 (C-34), 10.5 (C-38), -15.3 (C-41), 22.2 (C-42), 7.3 (C-44),
2(5)	59.8 (C-4)	22.9 (C-2), -15.5 (C-3), -6 (C-5), -6.8 (C-9), 10.5 (C-10), -16.7 (C-12), 23.7 (C-13), -6 (C-14), 8 (C-15), -6.7 (C-39), 10.8 (C-40),
37 (44)	53.8 (C-38)	9.2 (C-16), 6.9 (C-18), -6.1 (C-26), 22 (C-27), -15.3 (C-28), 10.4 (C-29), 7.1 (C-30), 7 (C-36), -6 (C-37), -14.9 (C-39), 23 (C-40),
58	56.3 (C-29)	6.6 (C-3), 6.6 (C-4), 9.9 (C-9), -6.2 (C-10), -15 (C-11), -15.1 (C-12), 21.6 (C-13), -6.3 (C-14), 10.1 (C-15), 6 (C-20), 6.1 (C-26), 6.5 (C-40), 6.6 (C-41), -6.2 (C-58), 21.2 (C-59)
11 (12)	52.7 (C-59)	-9.1 (C-2), 14.4 (C-3), -6.5 (C-8), 22.5 (C-9), -15.7 (C-10), -6.5 (C-11), 9.7 (C-20), 6.4 (C-29), 9.9 (C-33), -14.1 (C-42),
18	72 (C-17)	-6 (C-6), -6.8 (C-14), 8.6 (C-15), -22.1 (C-16), -7.5 (C-18), 11.3 (C-24), -7.7 (C-25), 31.2 (C-26), -9.9 (C-27), 14.7 (C-28), 6.5 (C-37)
28	55.4 (C-27)	6.3 (C-7), 6.4 (C-14), 19.9 (C-16), -8.6 (C-17), 14.8 (C-18), 7.3 (C-25), -15.6 (C-26), -6 (C-28), 21.1 (C-36), 7

Site of addition	Principal spin density and location	Spin densities (>6%) at other sites
		(C-38), 9.9 (C-48), -14.9 (C-49),
23 (29)	56.8 (C-52)	6.3 (C-22), -6.5 (C-23), 6.5 (C-24), 10.2 (C-36), 10 (C-45), -6.1 (C-49), 21.4 (C-50), -15 (C-51), -15.2 (C-53), 21.4 (C-54), -6 (C-55),
47 (31)	54.8 (C-46)	8.3 (C-31), -14.4 (C-34), -15.6 (C-45), -6 (C-47), 9.6 (C-53), -6 (C-54), 22.6 (C-55), -6.7 (C-56), 11.2 (C-57)

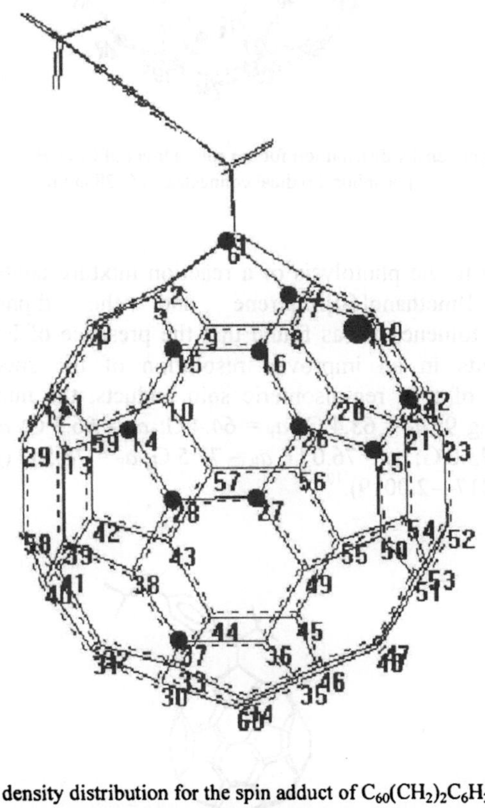

Figure 8. Spin density distribution for the spin adduct of $C_{60}(CH_2)_2C_6H_2(OMe)_2$ and phosphoryl radical connected to C-18 atom.

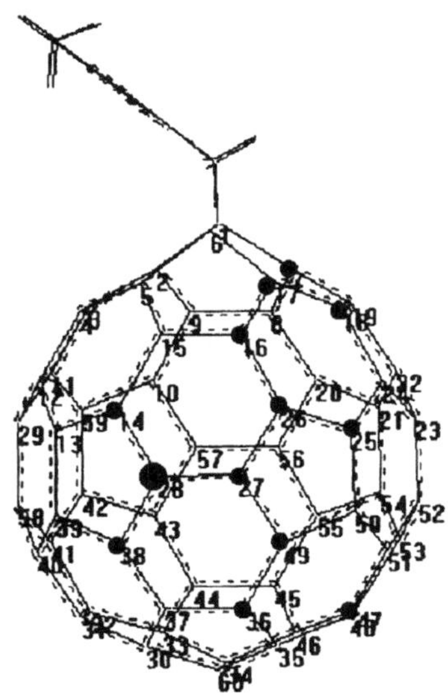

Figure 9. Spin density distribution for the spin-adduct of $C_{60}(CH_2)_2C_6H_2(OMe)_2$ and phosphoryl radical connected to C-28 atom.

In relation to the photolysis of a reaction mixture containing methyl(*p*-methoxyphenyl)methano[60]fullerene and the diphosphorylmercury compound in toluene, it was found that the presence of $HOC(CF_3)_3$ in the solution results in an improved resolution of the spectral pattern of superposition of the regioisomeric spin adducts, the number of isomers observed being 9: a_P = 63.4 G; a_P = 64.4 G; a_P = 66.3 G; a_P = 68.6 G; a_P = 70.5 G; a_P = 73.0 G; a_P = 76.0 G; a_P = 77.5 G; a_P = 78.5 G (*g*-factor for each isomer is 2.0017—2.0019).[10]

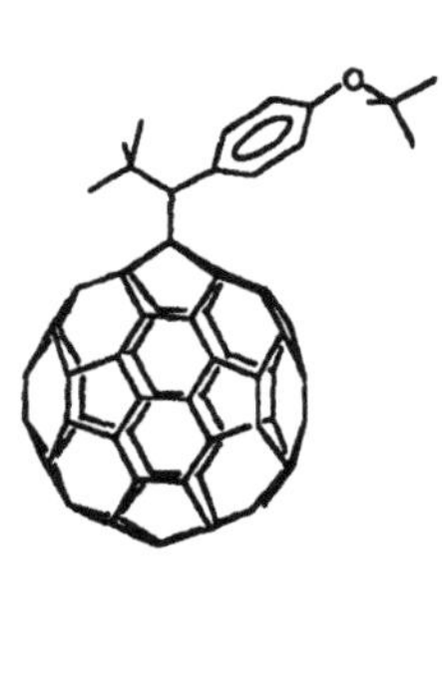

When irradiation has been switched off, regioisomeric spin adducts dimerize over a period of 1 s, except for three isomers, whose intensity decreases by 76, 96 and 87%, respectively. Evidently, the enthalpy of the radical—dimer equilibrium for these spin adducts decreases due to steric shielding by the methylene fragment; the integrated intensity of these signals is ~83% (Fig. 10).

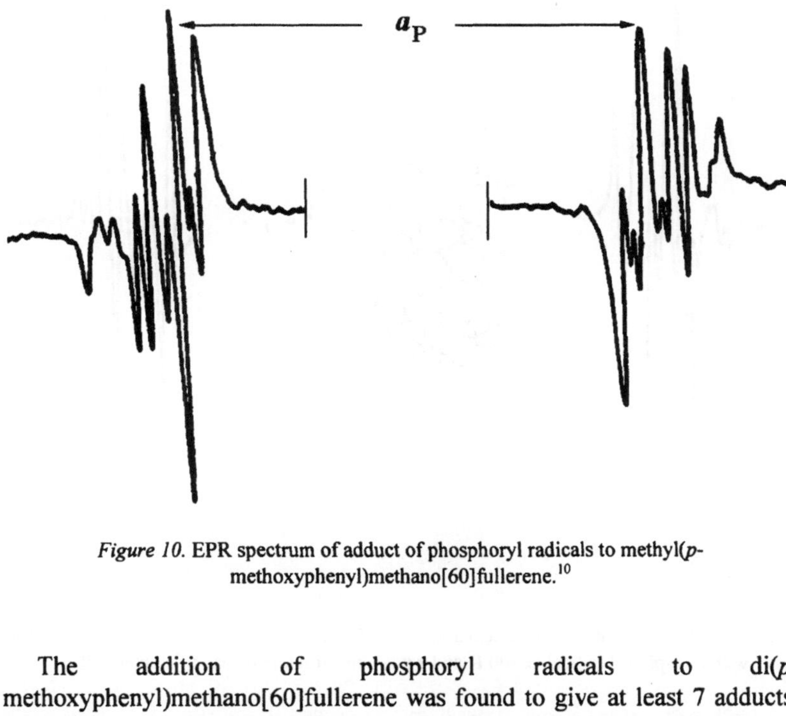

Figure 10. EPR spectrum of adduct of phosphoryl radicals to methyl(*p*-methoxyphenyl)methano[60]fullerene.[10]

The addition of phosphoryl radicals to di(*p*-methoxyphenyl)methano[60]fullerene was found to give at least 7 adducts, namely a_P = 74.4 G, g = 2.0024; a_P = 72.0 G, g = 2.0025; a_P = 68.5 G, g = 2.0023 (40%); a_P = 66.5 G, g = 2.0023; a_P = 64.8 G, g = 2.0020; a_P = 62.5 G, g = 2.0022; a_P = 59.25 G, g = 2.0021 (Fig. 11).[11,12]

All the radicals except for one vanish within 1 s (* in Fig. 11). Hence, this isomer has a lower enthalpy of equilibrium than the other isomers. As noted above, the reason is the close arrangement of the methano fragment to the radical center; *i.e.* in this isomer, the phosphoryl group is located most closely to the substituent.

An increase in temperature to 430 K results in dissociation of the dimers formed; in this case, the spectrum can be recorded without irradiation.

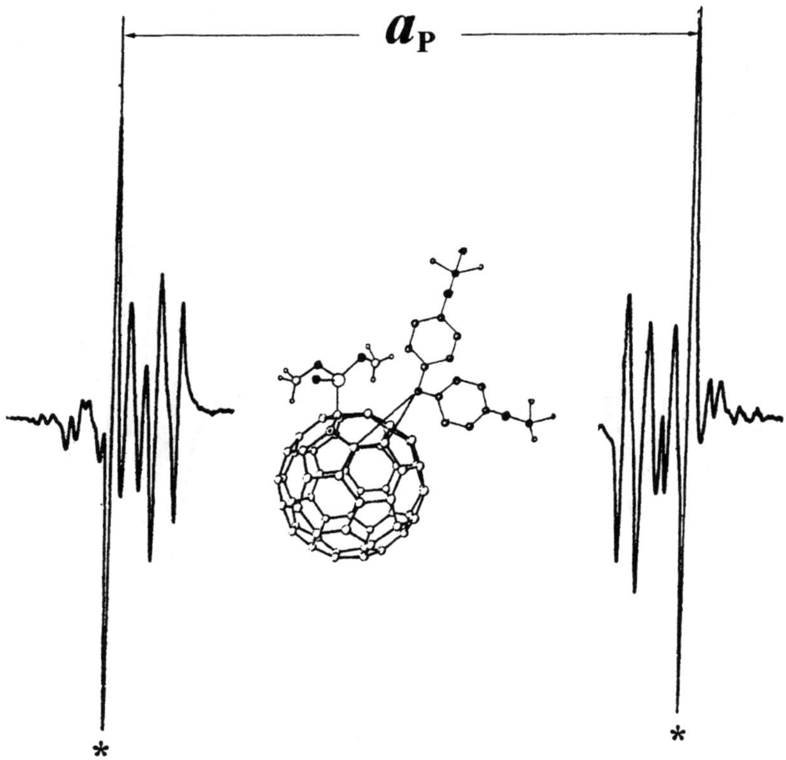

Figure 11. EPR spectrum of the radical adducts of di(*p*-methoxyphenyl)methano[60]fullerene with phosphoryl radical at 300 K and a possible structure of one of isomers (*).[11]

A different situation is observed when phosphoryl radicals undergo addition to spiro(2,6-di-*tert*-butyl-2,5-cyclohexadienone)-methano[60]fullerene.[10]

In this case, three types of species are detected, namely, three isomeric phosphorylfullerenyl radicals with the following parameters: a_P = 62.25 G, g = 2.0022; a_P = 64.0 G, g = 2.0020; and a_P = 66.0 G, g = 2.0023; a phenoxyl radical (a_H(2H) = 1.9 G, g = 2.0046); and a fullerenyl radical (singlet, g = 2.0024). The first three species are formed upon the attack by phosphoryl radicals on the fullerene-cage carbon atoms remote from the methylene group.

Photolysis of the reaction mixture results in photoexcitation of the starting compound giving rise to a biradical. The reaction of the phosphoryl radical with the fullerenyl biradical results in the formation of a stable fullerenylphenoxyl radical; this route of the attack may prove predominant:[10]

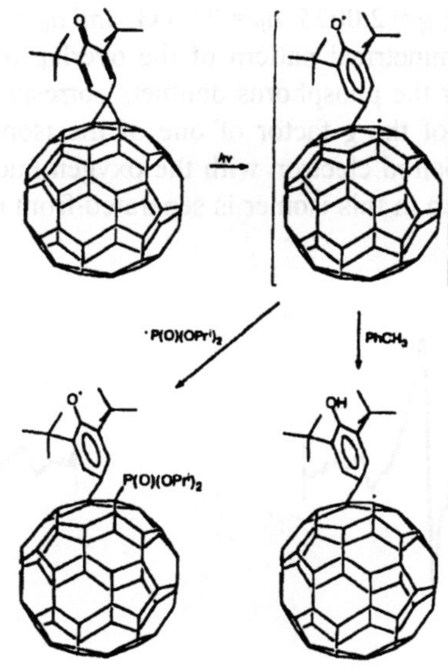

The spin adducts formed by phosphoryl radicals with the products of cycloaddition of substituted nitrile oxides to C_{60} (A and B) have also been studied by EPR:[13]

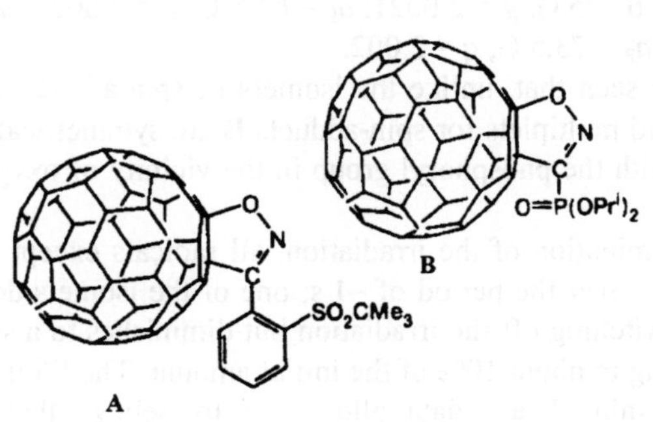

A

The addition of the phosphoryl radicals generated by the photolysis of $Hg[P(O)(OPr^i)_2]_2$ in a saturated toluene solution of compound **A** affords a mixture of isomers, in which the phosphoryl group is located at different distances from the organic substituent. The isomers differ in not only the *hfc* constants but also *g*-factors: a_P = 52.75 G, g = 2.0021; a_P = 62.25 G, g = 2.0021; a_P = 66.5 G, g = 2.0025; a_P = 72.5 G; and a_P = 75.0 G (Fig. 12). This results in a non-symmetrical pattern of the overlap of high-field and low-field components of the phosphorus doublets corresponding to some of the isomers. The shift of the *g*-factor of one of the isomers might be due to coupling of the unpaired electron with the oxygen nucleus. In other words, the phosphoryl group in this isomer is separated from the oxygen atom by 2 or 3 bonds.

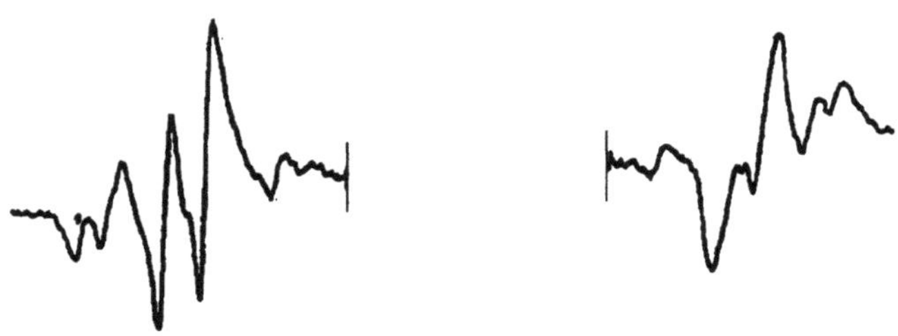

Figure 12. EPR spectrum of radical adducts of A and phosphoryl radical at 300 K.[13]

Photolysis of the diphosphorylmercury derivative in a saturated toluene solution of compound **B** yields a mixture of isomers having the following magnetic-resonance parameters: a_P = 52.9 G, g = 2.0022; a_P = 58.25 G, g = 2.0022; a_P = 62.25 G, g = 2.0021; a_P = 64.5 G, g = 2.0022; a_P = 67.25 G, g = 2.0021; and a_P = 73.5 G, g = 2.002.

It can be seen that, unlike the isomers of spin-adducts **A**, the low-field and high-field multiplets for spin-adducts **B** are symmetrical; hence, there is no isomer with the phosphoryl group in the vicinity of oxygen among these adducts.

Upon termination of the irradiation, all radicals except for one of them are destroyed over the period of ~1 s; one of the isomers does not decay to zero after switching off the irradiation but diminishes to a steady-state level corresponding to about 10% of the initial amount. The lifetime of this signal is about 1 min. These data allow one to believe that this radical is characterized by the closest attachment of phosphoryl to the heterocycle (the distance may be 2 or 3 bonds from the carbon atom).

EPR spectrum for the products of radical phosphorylation of $C_{60}Ph_5Cl$ show signals from four isomeric spin adducts (Fig. 13) (a_P = 88.7 G; a_P = 84.9 G; a_P = 78.7 G; a_P = 77.0 G).[14] The central signal in the spectrum is assigned to a stable fullerenyl radical $^{\bullet}C_{60}Ph_5$ resulting from homolytic cleavage of C–Cl bond by irradiation.[15]

As distinct from radical adducts of methanofullerenes, the radical adducts of $C_{60}Ph_5Cl$ have *hfc* constants with phosphorus nucleus 10–15 G higher, with 4 isomers forming. One can suppose the growth of the *hfc* constants to be associated with the presence of surface sites with lower extent of unpaired electron delocalization in $C_{60}Ph_5Cl$ molecule. Such a correlation between *hfc* constants with phosphorus nucleus and unpaired electron delocalization is clearly seen for the radical adducts of phosphoryl radicals with fullerenes C_{70} and C_{76}.[1] The decrease in the number of $^{\bullet}C_{60}Ph_5Cl[P(O)(OPr^i)_2]$ isomers may be due to higher symmetry of the molecule as compared with the symmetry of other studied fullerene derivatives.

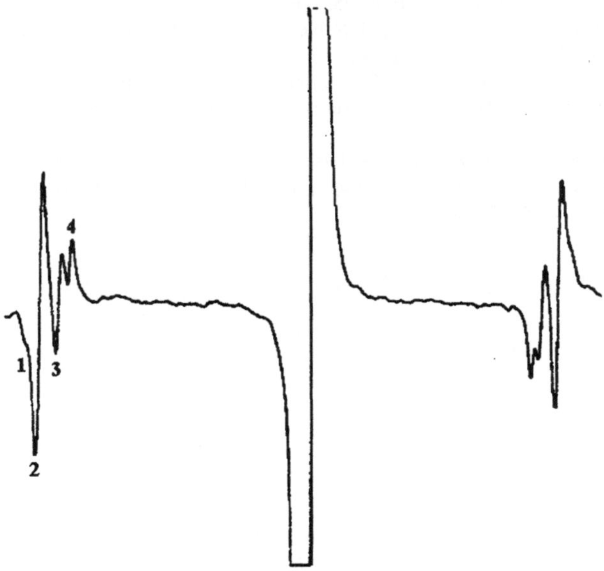

Figure 13. EPR spectra of the radical adducts of $C_{60}Ph_5Cl$ with phosphoryl radical and $^{\bullet}C_{60}Ph_5$.[14]

References

1. B. L. Tumanskii, V. V. Bashilov, O. G. Kalina, *Chemistry Reviews*, 2001, in press.
2. V. V. Bashilov, P. V. Petrovskii, V. I. Sokolov, S. V. Lindeman, I. A. Guzey, Yu. T. Struchkov, *Organometallics*, 1993, **12**, 991.
3. A. V. Usatov, K. N. Kudin, E. V. Vorontsov, L. E. Vinogradova, Yu. N. Novikov, *J. Organomet. Chem.*, 1996, **522**, 147.
4. B. L. Tumanskii, V. V. Bashilov, O. G. Kalina, V. I. Sokolov, *J. Organomet. Chem.*, 2000, **599**, 28.
5. B. L. Tumanskii, V. V. Bashilov, N. N. Bubnov, S. P. Solodovnikov, V. I. Sokolov, *Russ. Chem. Bull.*, 1994, **43**, 884.
6. B. L. Tumanskii, A. V. Usatov, V. V. Bashilov, S. P. Solodovnikov, N. N. Bubnov, Yu. N. Novikov, V. I. Sokolov, *Russ. Chem. Bull.*, 1997, **46**, 836.
7. B. L. Tumanskii, O. G. Kalina, V. V. Bashilov, A. V. Usatov, E. A. Shilova, Yu. I. Lyakhovetsky, S. P. Solodovnikov, N. N. Bubnov, Yu. N. Novikov, A. S. Lobach, V. I. Sokolov, *Russ. Chem. Bull.*, 1999, **48**, 1108.
8. B. de La Vassiere, J. P. B. Sandall, P. W. Fowler, P. de Oliveira, R. V. Bensasson, *J. Chem. Soc., Perkin Trans. 2*, 2001, 821.
9. O. G. Kalina, *Ph.D. Dissertation*, A. N. Nesmeyanov Institute of Organoelement Compounds, Russian Academy of Sciences, Moscow, Russia, 2001.
10. B. L. Tumanskii, E. N. Shaposhnikova, O. G. Kalina, V. V. Bashilov, M. N. Nefedova, S. P. Solodovnikov, N. N. Bubnov, V. I. Sokolov, *Russ. Chem. Bull.*, 2000, **49**, 843.
11. B. L. Tumanskii, M. N. Nefedova, V. V. Bashilov, S. P. Solodovnikov, N. N. Bubnov, V. I. Sokolov, *Russ. Chem. Bull.*, 1996, **45**, 2865.
12. B. L. Tumanskii, V. V. Bashilov, M. N. Nefedova, N. N. Bubnov, S. P. Solodovnikov, V. I. Sokolov, *J. Mol. Mat.*, 1998, **11**, 91.
13. B. L. Tumanskii, V. V. Bashilov, S. P. Solodovnikov, N. N. Bubnov, O. G. Sinyashin, I. P. Romanova, V. N. Drozd, V. N. Knyazev, V. I. Sokolov, *Russ. Chem. Bull.*, 1999, **48**, 1786.
14. B. L. Tumanskii, O. G. Kalina, V. V. Bashilov, R. Taylor, unpublished results.
15. R. G. Gasanov, O. G. Kalina, A. A. Popov, P. A. Dorozhko, B. L. Tumanskii, *Russ. Chem. Bull.*, 2000, **49**, 753.

Chapter 6

Paramagnetic Fullerene Derivatives with an Unpaired Electron on the Added Moiety

Homolytic reactions of fullerenes C_{60} and C_{70} are well studied. Carbon- and element-centered radicals have been reported to undergo regioselective addition to these fullerenes. The behavior of fullerene derivatives in radical processes remains much poorly understood. All of the homolytic reactions of the fullerene derivatives studied yield fullerenyl radicals with an unpaired electron on the carbon polyhedron.

Spin labeled fullerene derivatives observable by EPR are of special interest due to their use in biology and medicine for monitoring *in vivo* processes involving these bulky molecules capable of anchoring different functional and bioactive groups.[1]

Besides, the discovery of ferromagnetism for a complex of fullerene C_{60} with tetra(dimethylamino)ethylene, C_{60}(TDAE),[2] has aroused interest in fullerene derivatives whose unpaired electrons are localized both on the fullerene framework and on the attached group of atoms.

To obtain such compounds, nitroxyl and benzoquinone derivatives of fullerene were synthesized, which are potential synthons to study a possibility of dipole–dipole interaction of several unpaired electrons within one molecule.[3-5]

1.1 Nitroxyl Fullerene Derivatives

A series of C_{60} derivatives containing C_{2v}-symmetrical 2,2,5,5-teramethylpyrrolidine-1-oxyl have been synthesized.[3] A detailed EPR and ENDOR study of the neutral radicals and their reduction products were reported. All the neutral species gave EPR spectra consisting of a main triplet with ~15 G splitting by the ^{14}N nucleus, typical of nitroxyl radicals (Fig. 1a).

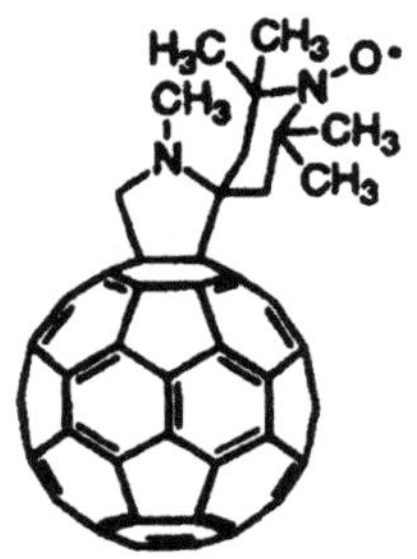

I

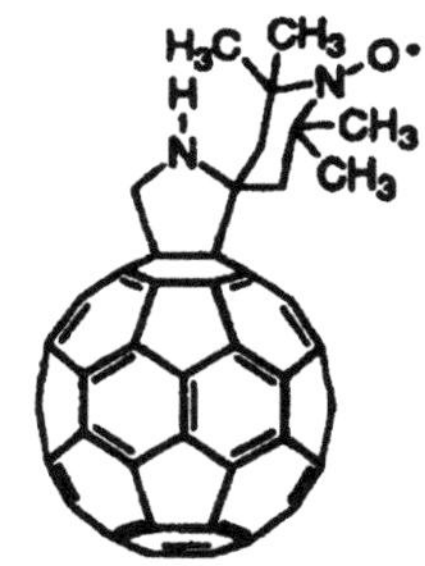

II

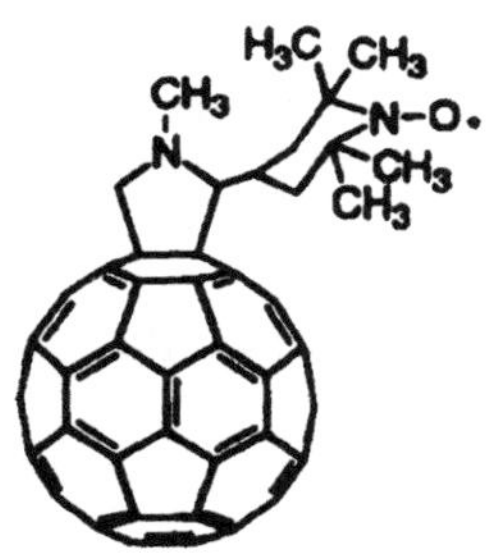

III

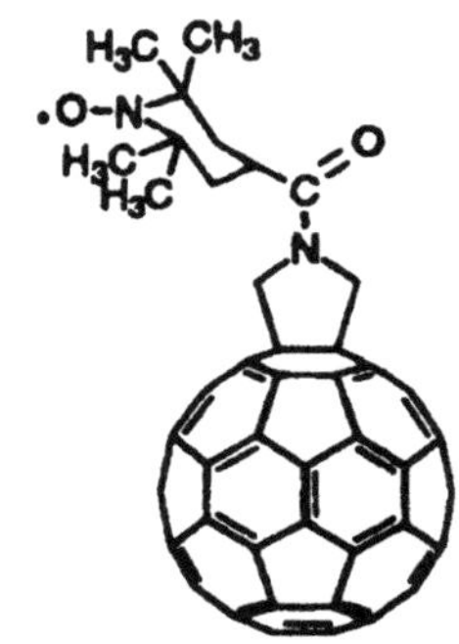

IV

Preparation of I. *A solution of 100 mg (0.14 mmol) of C_{60}, 97 mg (0.6 mmol) of 4-piperidone-1-oxyl, and 87 mg (0.98 mmol) of N-methylglycine in 100 mL of toluene was stirred at reflux temperature for 7 h, and then the solvent was removed in vacuo. The residue was purified by flash column chromatography (toluene as eluent), affording 16 mg (13%) of the nytroxyl fullerene derivative with 80 mg (80%) of unreacted C_{60}.*[3]

When the compounds were gradually reduced under vacuum by exposure to an alkali metal mirror, new EPR lines were observed; a 1 : 1 : 1 triplet with a splitting constant of one-half that of a nitroxyl radical and g = 2.0030 is attributed to biradical anions where one electron is located in the fullerene moiety and experiences a strong exchange coupling with the nitroxyl unpaired electron (Fig. 1b) (Table 1).[3]

Table 1. g-Factors and *hfc* constants for fullerenyl nitroxyl derivatives.

Compound	state	g-factor	*a*, G
I	neutral	2.0061	15.30
	mono-anion	2.0030	7.65
II	n	2.0061	15.15
	ma	2.0030	7.80
III	n	2.0061	15.45
	ma	2.0030	7.83
IV	n	2.0061	15.38
	ma	2.0030	7.57

Frozen solution spectra of these compounds allowed the determination of the electron-electron dipolar interaction parameters. The experimental values agree with a spin distribution of one unpaired electron mostly in the equatorial plane of the fullerene. Another single line, which appeared in the spectrum as the reduction continues to proceed (g = 1.9999), was attributed to a new species: a substituted fullerene radical anion in which the nitroxyl group was irreversibly reduced (Fig. 1b).[3]

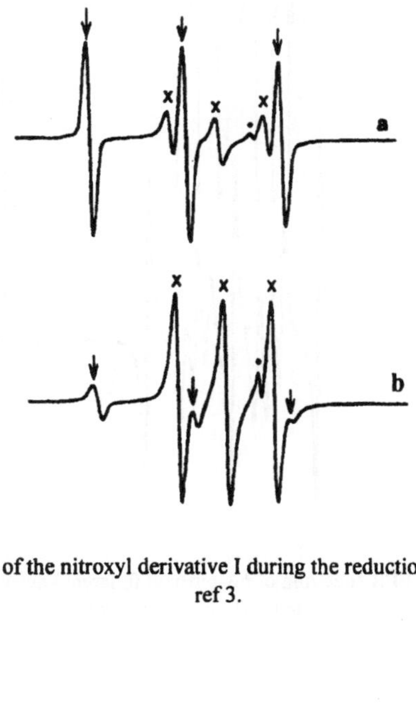

Figure 1. EPR spectra of the nitroxyl derivative I during the reduction process. Adapted from ref 3.

For fullerene linked to a nitroxyl radical, a time-resolved EPR spectrum was reported in toluene solution and EPR signal of the excited quartet in toluene solution was assigned on the basis of hyperfine coupling constants and g-factors.[6]

[60]Fullerene nitroxyl bis(adducts) have been synthesized.[7,8]

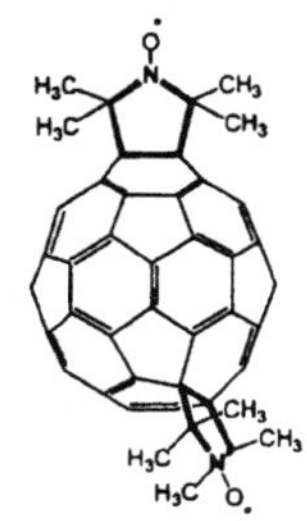

Their EPR spectrum is typical of nitroxyl biradicals with electron exchange interaction **J** of the same order of the nitrogen hyperfine coupling. The absolute value of **J** was obtained from the EPR spectrum, while the sign was measured by ENDOR. Strong spin polarization was observed when the derivatives were photoexcited. This behavior was accounted for by an intramolecular triplet-triplet interaction (Fig. 2).[7]

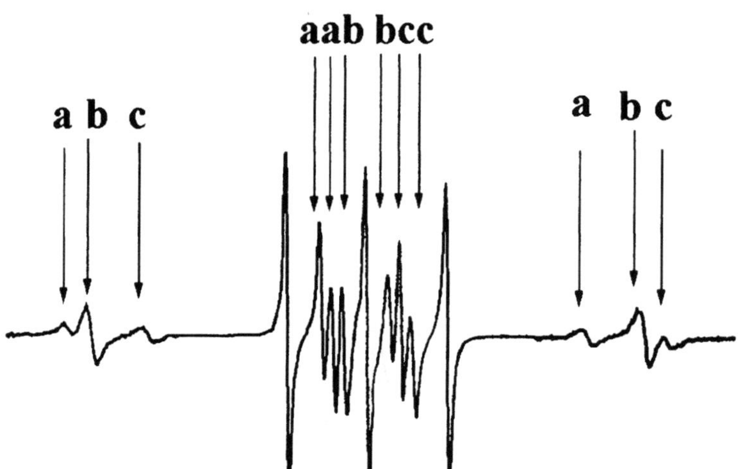

Figure 2. Steady-state EPR spectrum of *bis*-nitroxyl fullerenyl derivative. a, b, c labels indicate 3 AB systems corresponding to the different ^{14}N configurations. Adapted from ref 7.

A 2D nutation spectrum for radical-triplet pair system of fullerene linked to a nitroxyl radical was observed and then nutation frequencies of the signals were determined and multiplicities of the involved states were assigned.[8]

1.2 Benzoquinone Fullerene Derivatives

Another approach to the structures with several unpaired electrons within one molecule consists in preparing benzoquinone derivatives of C_{60} where benzoquinone moiety could be transformed into semiquinonato fragment via radical addition or reduction with alkali metals.

The synthesis and some properties of the C_{60} derivative covalently linked to benzoquinones was reported.[4,5]

O-Benzoquinonefullerene[5] was prepared by oxidizing a solution of corresponding catechol in benzene with an excess of Ag_2O. Similar *p*-benzoquinone derivatives were obtained by oxidation of the corresponding catechol with DDQ.[4]

The formation of paramagnetic metal chelates containing fullerene fragment was detected for the first time by EPR in the reactions of Mn- and Re-centered radicals with the *o*-benzoquinone derivative of fullerene C_{60}.[5]

After $Mn_2(CO)_{10}$ or $Re_2(CO)_{10}$ in an equimolar amount was added, the reaction mixture was irradiated in a resonator of an EPR spectrometer. As a result, the reaction of metal-centered radicals with compound **A** yields stable paramagnetic complexes:

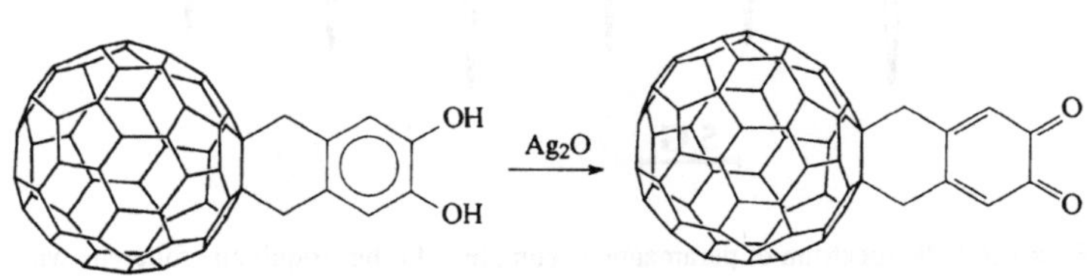

A

$$M_2(CO)_{10} \xrightarrow{h\nu} 2\,{}^{\bullet}M(CO)_5 \quad (M = Mn, Re)$$

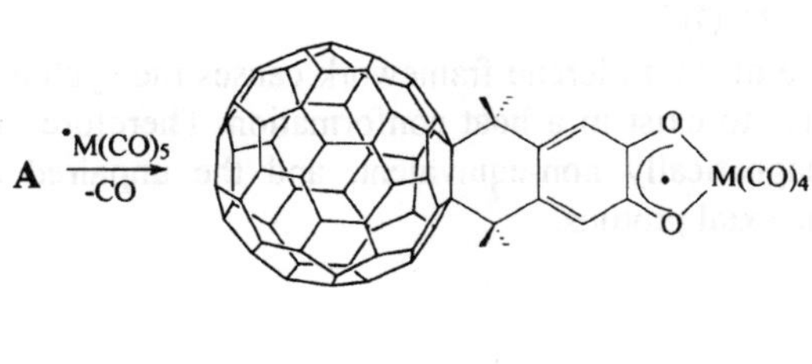

Preparation of *o*-benzoquinonefullerene[60]. *A tenfold excess of Ag_2O was added to a 0.001 M solution of catechol in benzene. The reaction mixture was stirred for 40 min. The precipitate formed was filtered off, and the filtrate was concentrated. Trituration of the residue with hexane gave compound A in 97% yield as a finely crystalline dark brown powder.* [5]

The EPR spectrum of the reaction mixture shows a sextet of triplets (Fig. 3, 4). Such a picture is attributed to an interaction of the unpaired electron with the ^{55}Mn or $^{185,187}Re$ (I = 5/2) and with two protons of the methylene groups that are axial to the plane of the aromatic ring (a_{Mn} = 7.5 G, a_H (2H) = 6 G, g= 2.0034; a_{Re} = 30 G, a_H = 6 G, g = 2.0034).

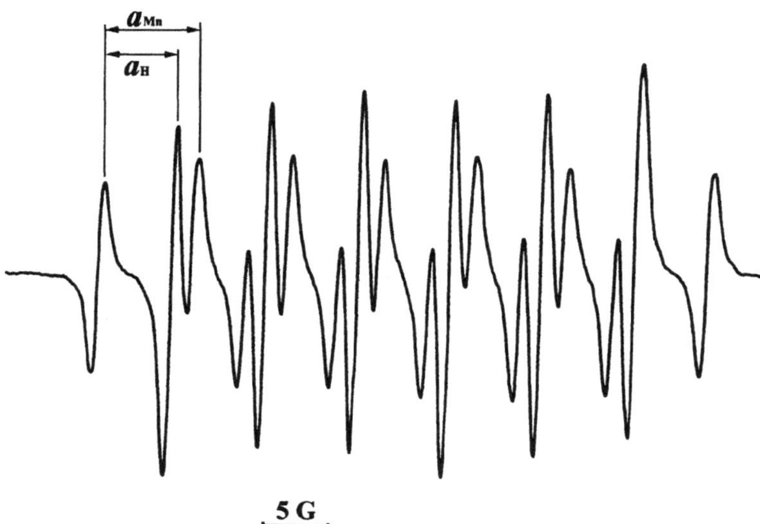

Figure 3. EPR spectrum of paramagnetic complex of *o*-benzoquinone fullerene with $Mn(CO)_4$.[5]

When PPh_3 was added, the CO ligands in the metal coordination sphere were replaced to give disubstituted complexes (EPR: a_H(2 H) = 5.5 G, a_p(2 P)= 16.5G, a_{Mn}= 11 G).[5]

The presence of the fullerene framework causes the cyclohexane ring in these compounds to exist in a boat conformation. Therefore the methylene protons are magnetically non-equivalent, and the unpaired electron can interact with the axial protons.

Figure 4. EPR spectrum of paramagnetic complex of *o*-benzoquinone fullerene with $Re(CO)_4$.[5]

Reduction of benzoquinone-linked fullerene with alkali metals and by CVA has been investigated and the corresponding anion radicals have been identified (Table 2).[4]

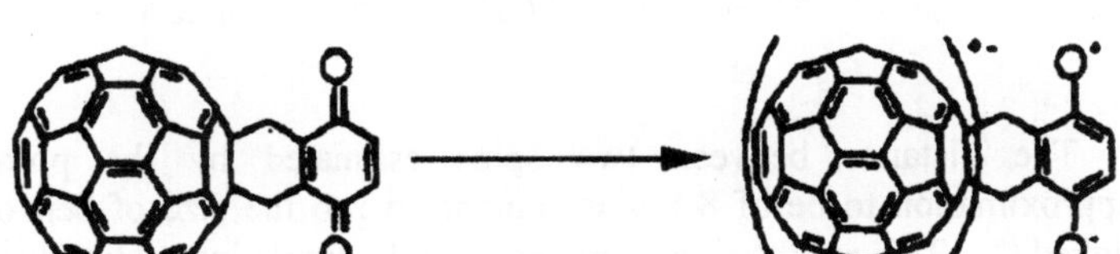

Table 2. Redox potentials for the *p*-QC_{60} derivative.[4]

Compound	E_{red}^{1} (quinone)	E_{red}^{1} (C_{60})	E_{red}^{2} (C_{60})	E_{red}^{2} (quinone)	E_{red}^{3} (C_{60})
C_{60}		-1.12	-1.52		-1.98
p-QC_{60}	-1.02	-1.22	-1.63	-1.99	-2.28

Reduction of benzoquinone-linked C_{60} with potassium in THF gives the EPR spectra of the dianion, which consists of a broad signal at the center and a fine structure due to triplet species $|D|$ = 45.8 G at 77K. The sodium reduction of benzoquinone-linked C_{60} in THF gives almost the same results as potassium $|D|$ = 46.0 G at 77K (Fig. 5).[4]

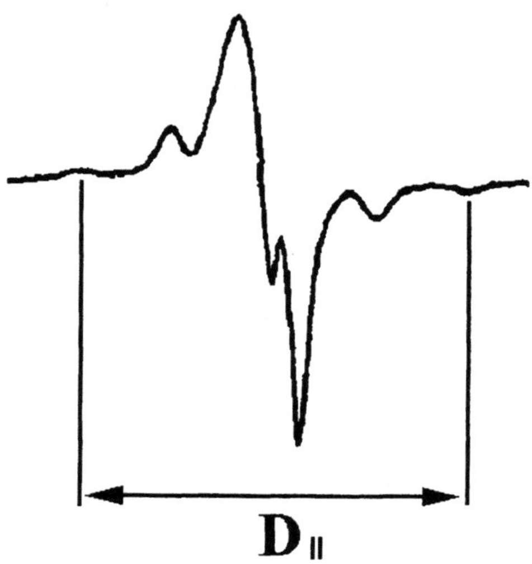

Figure 5. EPR spectrum of *p*-$QC_{60}^{2-}Na_2^{+}$ at 77 K. Adapted from ref 4.

The distance between two spins estimated by the point dipole approximation to be of 8.5 A is comparative to the size of benzoquinone-linked C_{60}. This results show a marked dipole-dipole interaction between the fullerene core and its substituent at the ground state.[4]

Stepwise reduction of *o*-QC_{60} with sodium amalgam has been found. First, *o*-benzoquinone fragment and then the fullerene core has been reduced, as evidenced by the decrease in the *hfc* constant with protons of methylene groups presumably situated in the axial positions about the plane perpendicular to $2p_z$-orbital of the unpaired electron.[9]

The reduction with potassium amalgam leads to fullerene-containing semiquinonato derivative in which unpaired electron interacts with all hydrogens of methylene groups and benzene ring (Fig. 6).[9]

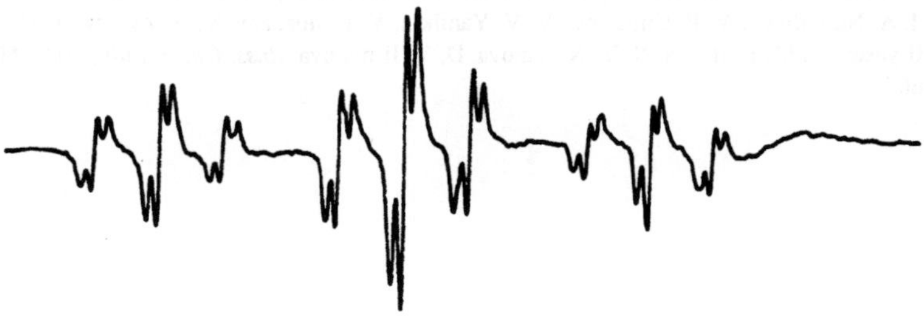

Figure 6. EPR spectrum of *o*-$QC_{60}^{-}K^{+}$.[9]

1.3 Phenoxyl Fullerene Derivatives

A phenoxyl radical was obtained by chemical oxidation of fulleropyrrolidine containing a sterically hindered phenolic fragment with PbO_2 (g = 2.0045, a_{2H} = 1.8 G, a_H = 4.7 G, a_N = 0.9 G).[10]

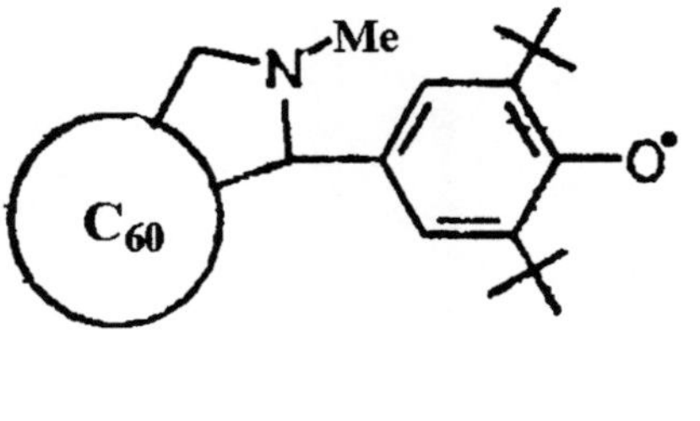

References
1. *Spin Labeling*, Ed. by L. J. Berliner, Academic Press, New York, 1976.
2. P.-M. Allemand, K. C. Khemani, A. Koch, F. Wudl, K. Holczer, S. Donovan, G. Gruner, J. D. Tompson, *Science*, 1991, **253**, 301.
3. F. Arena, F. Bullo, F. Conti, C. Corvaja, M. Maggini, M. Prato, G. Scorrano, *J. Am. Chem. Soc.*, 1997, **119**, 789.
4. M. Iyoda, S. Sasaki, F. Sultana, M. Yoshida, Y. Kuwatani, S. Nagase, *Tetr. Lett.*, 1996, **37**, 7987.

5. O. G. Kalina, V. V. Bashilov, A. A. Khodak, B. L. Tumanskii, *Russ. Chem. Bull.*, 2001, **50**, 566.
6. C. Corvaja, M. Maggini, M. Prato, G. Scorrano, M. Venzin, *J. Am. Chem. Soc.* 1995, **117**, 8857.
7. C. Corvaja, L. Franco, M. Mazzoni, M. Maggini, G. Zordan, E. Menna, G. Scorrano, *Chem. Phys. Lett.*, 2000, **330**, 287.
8. N. Mizuochi, Y. Ohba, S. Yamauchi, *J. Phys. Chem. A*, 1999, **103**, 7749.
9. O. G. Kalina, *Ph.D. Dissertation*, A. N. Nesmeyanov Institute of Organoelement Compounds, Russian Academy of Sciences, Moscow, Russia, 2001 (for Sophia).
10. I. A. Nuretdinov, V. P. Gubskaya, V. V. Yanilkin, V. I. Morozov, V. V. Zverev, A. V. Il'yasov, G. M. Fazleeva, N. V. Nastapova, D. V. Il'matova, *Russ. Chem. Bull.*, 2001, **50**, 607.

Chapter 7

Radical Addition of Macromolecules to C_{60}

Macromolecular fullerene derivatives facilitate the combination of the unusual properties of fullerenes with the specific properties of polymers. Thus, macromolecular modification of fullerenes can provide polymeric materials with interesting properties and technological applications.[1,2]

It is not our purpose to analyze how the structure of fullerene-containing macromolecules affects their properties. We will discuss different methods used for the synthesis of polymers by radical polymerization.

There are three main approaches to the polymerization in the presence of C_{60}. The first one is the use of free radical initiators for copolymerization of C_{60} with vinyl monomers such as styrene and methyl methacrylate.[3-5] The second one is the nitroxyl-mediated "living" radical polymerization (NMRP).[2,6] And the third one is the atom-transfer radical polymerization (ATRP).[7]

The second and the third techniques employ the principle of an equilibrium between a low concentration of active and a rather large number of dormant compounds. This suppresses bimolecular side reactions such as recombination or disproportionation to such a degree that the overall polymerization process shows the increase of the molecular weight with conversion and the possibility to synthesize block copolymers.

1.1 Radical Polymerization in the Presence of Free Radical Initiators

This method is the easiest for direct incorporation of C_{60} into polymer by free-radical copolymerization under routine conditions. With AIBN as an initiator of free-radical polymerization of styrene (S) in the presence of C_{60}, the following mechanism was suggested:[3,8]

$$AIBN + styrene \longrightarrow PS^{\bullet} \quad (1)$$

$$PS^{\bullet} + C_{60} \longrightarrow PSC_{60}^{\bullet} \quad (2)$$

$$PSC_{60}^{\bullet} + S \longrightarrow PS\text{-}C_{60}\text{-} PS^{\bullet} \quad (3)$$

$$PSC_{60}^{\bullet} + PS'^{\bullet} \longrightarrow PS\text{-}C_{60}\text{-}PS' \quad (4)$$

There are no evidences to suppose that the fullerenyl radical can undergo addition to the double bond of the styrene monomer in reaction (3); if reaction (4) dominates, then there should be only one C_{60} per chain.

The yield of polymer produced by a solution polymerization is relatively low (ca. 20%) that suggests free radical is to be trapped on the fullerene (reaction 2) and does not propagate but can terminate (reaction 4).

Fullerene-styrene random copolymers with novel optical properties were obtained either under thermal initiation (heating above 130 °C) or in the presence of a radical initiator.[9] The copolymers from the two different methods have similar physical properties but different molecular weight distributions.

The effect of C_{60} on the radical polymerization of methyl methacrylate (MMA) in benzene has been studied kinetically and by EPR, with dimethyl-2,2'-azo-*bis*-(isobutyrate) (MAIB) being used as initiator.[4] The polymerization rate $\mathbf{R_p}$ and the molecular weight of resulting poly(MMA) decreased with increasing C_{60} concentration ((0-2.11) $\times$ 10^{-4} mol l^{-1}). The molecular weight of the polymer tended to increase with time at higher concentrations of C_{60}. The value of $\mathbf{R_p}$ at 60 °C in the presence of C_{60} (7.0 $\times$ 10^{-6} mol l^{-1}) was expressed by equation $R_p = k[MAIB]^{0.5}[MMA]^{1.25}$. The overall activation energy of polymerization at C_{60} concentration 7.0 $\times$ 10^{-6} mol l^{-1} was calculated to be 23.2 kcal mol^{-1}.

The reactivity of C_{60} toward propagating poly(MMA) radical was estimated by the method where the present polymerization is considered to include the following reactions:

$$MAIB \rightarrow 2R^{\bullet} \quad R_d = 2k_d\,[MAIB] \quad (5)$$

$$R^{\bullet} + MMA \rightarrow P^{\bullet} \quad R_i = k_i[R^{\bullet}][MMA] \quad (6)$$

$$P^{\bullet} + MMA \rightarrow P^{\bullet} \quad R_p = k_p[P^{\bullet}][MMA] \tag{7}$$

$$P^{\bullet} + C_{60} \rightarrow PC_{60}^{\bullet} \quad R_{C60} = k_{C60}[P^{\bullet}][C_{60}] \tag{8}$$

$$PC_{60}^{\bullet} + MMA \rightarrow P^{\bullet} \quad R_o = k_o[PX^{\bullet}][MMA] \tag{9}$$

$$PC_{60}^{\bullet} + P^{\bullet} \rightarrow \text{polymer} \qquad R_c = k_c\,[PC_{60}^{\bullet}]\,[P^{\bullet}] \tag{10}$$

$$PC_{60}^{\bullet} + PC_{60}^{\bullet} \rightarrow \text{polymer} \qquad R_z = k_z\,[PC_{60}^{\bullet}]^2 \tag{11}$$

$$P^{\bullet} + P^{\bullet} \rightarrow \text{polymer} \qquad R_t = k_t[P^{\bullet}]^2 \tag{12}$$

Here R is a primary radical from MAIB, $P^{\bullet}$ - propagating poly(MMA) radical. R_o, R_p, R_c, R_z, R_x, k_o, k_p, k_c, k_z, and k_t are the rates and rate constans of elementary reactions, respectively. The rate constant for the addition of poly(MMA) radical to C_{60} was estimated to be 6920 l mol^{-1} s^{-1} that means the high reactivity of the propagating poly(MMA) toward C_{60}.

Stable fullerenyl radicals were observed by EPR in the polymerization system. The concentration of fullerenyl radicals was found to increase linearly with time and then be saturated. The rate of fullerenyl radical formation increased with MAIB concentration. The $PC_{60}^{\bullet}$ concentration showed a tendency to decrease after saturation, indicating that some of the stable $PC_{60}^{\bullet}$ can react with newly formed poly(MMA) radicals. The analogous spectra were obtained for the radical polymerization of styrene in the presence of C_{60}.[4]

The thermal polymerization of styrene (S) in the presence of poly(MMA) containing C_{60} has been studied. It was found that C_{60} undergoes multiple additions of propagating polymer radicals. The MMA-S copolymer is considered to be formed by addition of PS radicals (generated by thermal initiation) to C_{60} containing poly(MMA). As a result, the copolymer may be a kind of a star polymer bearing poly(MMA) and PS arms located on the surface of C_{60}.[4]

The structures of C_{60}-styrene and C_{60}-methyl methacrylate copolymers were studied in details by W. T. Ford, et.al.[5] Polymerizaton in 1,2-dichlorobenzene solutions containing styrene or methyl methacrylate and monomer/C_{60}/azo(bis(isobutyronitrile)) leads to high molecular weight materials in which all of the C_{60} fragments are incorporated covalently. To understand the structures of the polymers and their mechanism of formation, the samples were isolated after low conversion of monomer and analyzed. Lower intrinsic viscosity and higher absolute molecular weight of the fullerene-containing polymers as compared with linear polystyrenes at equal retention time show that the polymer structures are branched.

Polymeric adducts of alkyl radicals with C_{60} also have been closely investigated. Polymeric radicals were generated via the photolysis of benzophenone (BP) and low-density polyethylene (LDPE) in the presence of C_{60} both in solution and in the molten state.[10]

EPR measurements and simulation allow one to assign the observed spectra and to reveal structural details for the polymeric alkyl-$C_{60}^{\bullet}$ adducts. UV irradiation of toluene solution of LDPE/BP/C_{60} made it possible to fix an intense and well-resolved EPR spectrum of fullerenyl radicals $PhCH_2C_{60}^{\bullet}$ and pairs of satellites due to ^{13}C hyperfine interactions (Fig. 1). The relative intensities and hyperfine couplings of these ^{13}C satellite transitions in natural abundance are consistent with the C_s symmetry of the radical adducts as in other alkylfullerenyl radicals in which the unpaired electron locates on the two fused six-membered rings of the C_{60} surface.

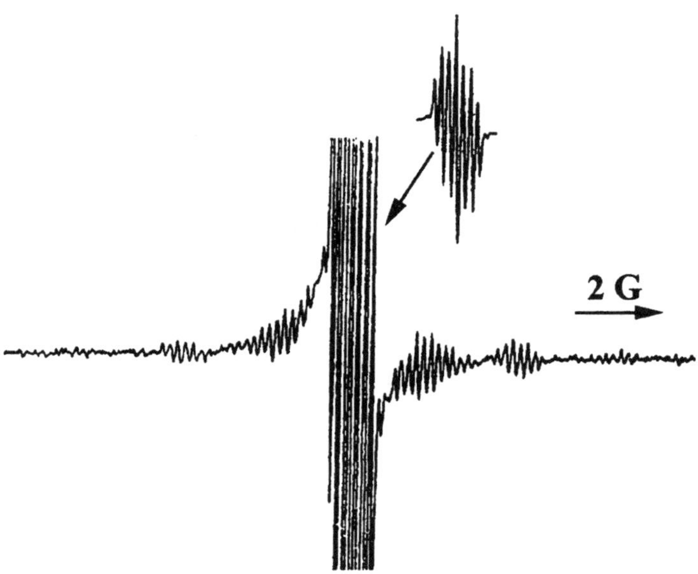

Figure 1. EPR spectrum of UV irradiated LDPE/BP/C_{60}. Adapted from ref 10.

Well-resolved proton coupling agrees well with data obtained for $PhCH_2C_{60}^{\bullet}$ produced via hydrogen atom abstraction by $^{\bullet}OC(CH_3)$ from toluene.

The *hfi* with protons disappears with increasing the temperature from 300 to 373 K.

Another EPR spectrum is recorded upon irradiation of LDPE/BP/C_{60} in benzene. As it can be seen from the Figure 2, the spectrum shows a 12-line hyperfine structure of signal overlapped with a broad peak. It was suggested, that polymeric alkyl radicals ($P^{\bullet}$) may be produced through hydrogen abstraction by the BP excited triplet state from LDPE chains followed by addition onto C_{60} to form PC_{60} radical adduct. The best result was obtained upon irradiation of LDPE/BP/C_{60} in the absence of the solvent.

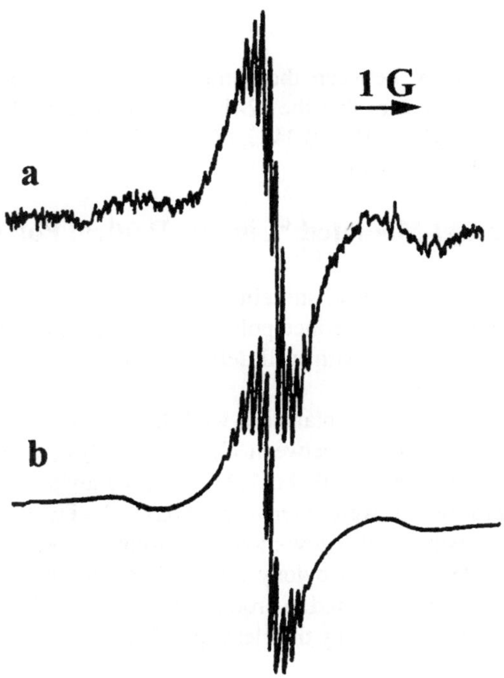

Figure 2. EPR spectrum of LDPE/BP/C_{60} after UV irradiation at 413 K (a) and its simulation (b). Adapted from ref 10.

According to the data obtained, the two kinds of polymeric alkyl radicals (i.e. tertiary and secondary carbon radicals) can be obtained upon irradiation of LDPE/BP.[10] These polymeric alkyl radicals could react with C_{60} to yield paramagnetic adducts $P_AC_{60}^{\bullet}$ and $P_BC_{60}^{\bullet}$:

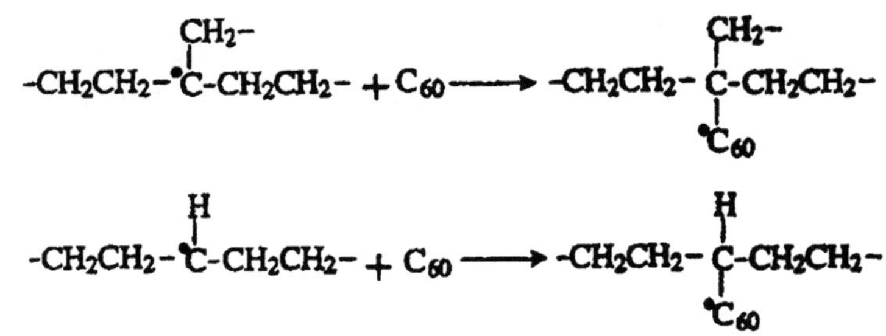

A good agreement between the simulated spectrum and the observed spectrum (Fig. 2) indicates that the 12-line hfs signal indeed arises from the overlapping of $P_AC_{60}^{\bullet}$ [a(3H) = 0.34 G, a(3H) = 0.17 G] and $P_BC_{60}^{\bullet}$ [a(1H) = 0.49, a(4H) = 0.13G] spectra.

1.2 Nitroxyl-Mediated "Living" Radical Polymerization

In the case of free radical initiation, the C_{60}-styrene copolymerization result in mixtures of poorly defined polyadducts, and it seemed very difficult to prepare PS-C_{60} adducts with well-defined structure by the usual radical polymerization methods.

H. Okamura et. al. have obtained a well-defined polymer derivatives of C_{60} by the addition reaction between C_{60} and a polystyryl radical derived from PS-TEMPO adduct (TEMPO is 2,2,6,6-tetramethylpiperidinyl-1-oxy).[6]

For this purpose, it were prepared several PS-TEMPO from PS of narrow-polydispersity and a low-molecular-weight model compound BS-TEMPO, where BS is a 2-benzoyloxy-1-phenylethyl group.

These adducts were expected to produce polystyryl (or benzoyloxystyryl) radicals at high temperatures by the cleavage of the C-ON bond:

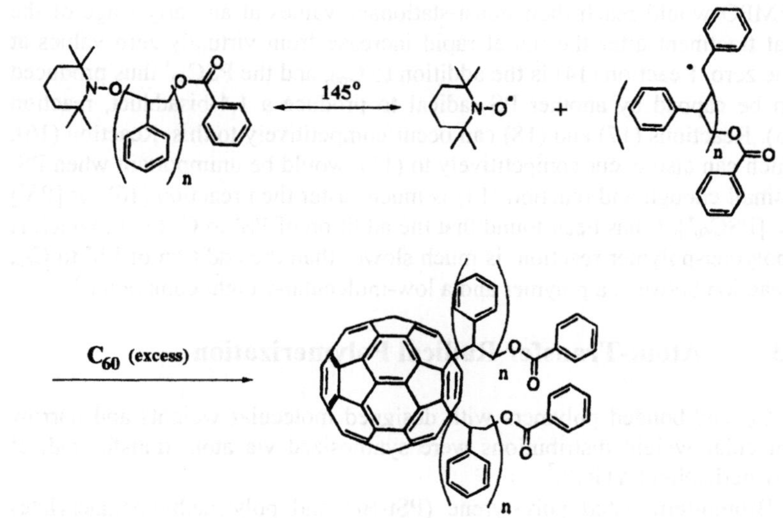

Possible reactions in a C_{60}/PS-TEMPO system at a high temperature can include the following transformations:[6]

$$\text{PS-TEMPO} \leftrightarrow \text{PS}^\bullet + \text{TEMPO}^\bullet \quad (13)$$

$$\text{PS}^\bullet + C_{60} \rightarrow \text{PSC}_{60}^\bullet \quad (14)$$

$$\text{PSC}_{60}^\bullet + \text{PS}^\bullet \rightarrow C_{60}(\text{PS})_2 \quad (15)$$

$$2\text{PS}^\bullet \rightarrow (\text{PS})_2 \ \text{(mainly)} \quad (16)$$

$$2\text{PSC}_{60}^\bullet \leftrightarrow \text{PSC}_{60}\text{-}C_{60}\text{PS} \quad (17)$$

$$\text{PSC}_{60}^\bullet + \text{TEMPO} \leftrightarrow \text{PSC}_{60}\text{-TEMPO} \quad (18)$$

Reaction (13) is a reversible dissociation of the polystyroxyamine, whose dissociation and association rate are known. The concentration of $\text{PS}^\bullet$ and

TEMPO would reach their quasi-stationary values at an early stage of the heat treatment after the initial rapid increase from virtually zero values at time zero. Reaction (14) is the addition to C_{60}, and the $PSC_{60}^{\bullet}$ thus produced can be capped by another PS radical to produce a 1,4-bisadduct, reaction (15). Reactions (17) and (18) can occur competitively to this. Reaction (16), which can also occur competitively to (15), would be unimportant when $PS^{\bullet}$ is small enough and reaction (14) is much faster then reaction (16), or $[PS^{\bullet}] \ll [PSC_{60}^{\bullet}]$. It has been found that the addition of $PS^{\bullet}$ to $C_{60}(PS)_2$, which is a polymer-polymer reaction, is much slower than the addition of $PS^{\bullet}$ to C_{60}, a reaction between a polymer and a low-molecular-weight compound.[6]

1.3 Atom-Transfer Radical Polymerization

C_{60}-end-bonded polymers with designed molecular weights and narrow molecular weight distributions were synthesized via atom transfer radical polymerization (ATRP).[7]

Bromo-terminated polystyrene (PSt-Br) and poly(methylmethacrylate) (PMMA-Br) with designed molecular weights and narrow molecular weight distributions were prepared by ATRP. These bromo-terminated polymers were allowed to react with C_{60} under ATRP conditions to yield C_{60}-terminated PSt (PSt-C_{60}-Br) and PMMA (PMMA-C_{60}-Br). The polymers were characterized by gel permeation chromatography (GPC), UV–vis, FT-IR, and fluorescence spectra. The mechanism of the preparation of C_{60}-terminated polymers is as follows:[7]

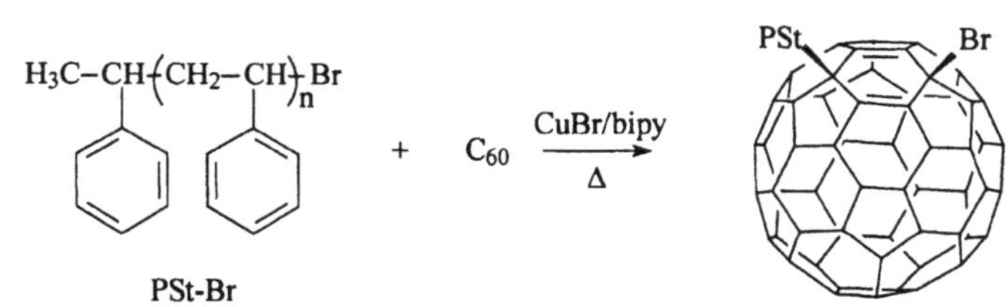

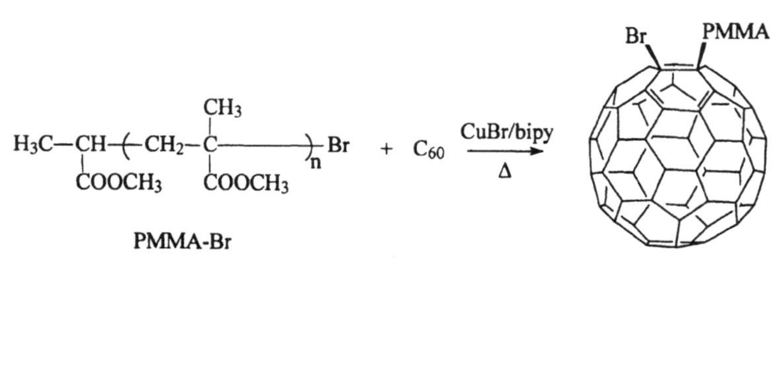

A catalytic amount of the CuBr/bipy coordination complex reversibly abstracts the bromine atoms from the polymer chain ends switching them from a dormant state to an active state and generates a small concentration of polymer-based radicals and the corresponding $CuBr_2$/bipy complex. Since the $CuBr_2$/bipy complex is only slightly soluble in chlorobenzene, the first reaction in this process is a heterogeneous one.

Thermodynamically, the equilibrium must lie toward the side of dormant chain end to maintain a sufficiently low concentration of $PSt^{\bullet}$, so the bimolecular termination between radicals was suppressed.

Kinetically, the first reaction is fast, thus keeping the concentration of radicals in a steady state.

The second reaction is the addition of radicals to C_{60} to produce $PStC_{60}^{\bullet}$. In the next reaction, $PStC_{60}^{\bullet}$ abstracts a bromine atom from the $CuBr_2$/bipy complex to yield the final product $PStC_{60}Br$.

At the same time, the terminations between the following radicals were unavoidable:[7]

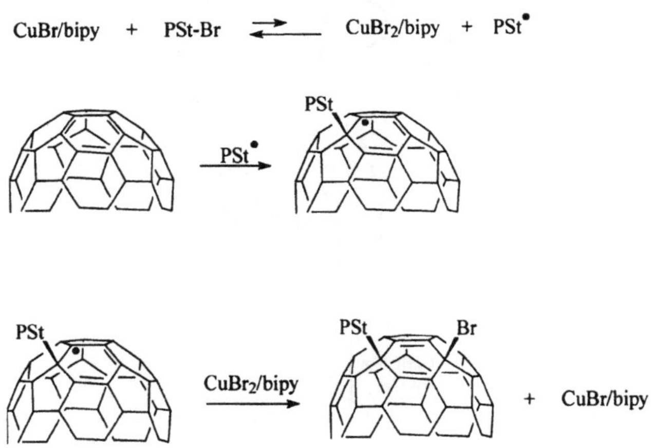

In the next studies, both methods of the regular radical polymerization were used: NMRP and ATRP.

A donor-acceptor, rod-coil diblock copolymer has been synthesized with the objective of enchancing the photovoltaic efficiency of the PPV-C_{60} (PPV = poly(*p*-phenylenevinylene) system by the incorporation of both components in molecular architecture that is self-structuring trough microphase separation:[2]

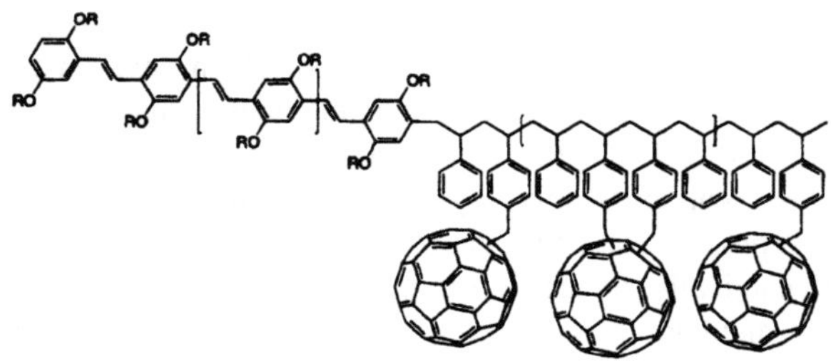

Diblock copolymers were obtained by using an end-functionalized rigid-rod block of poly(2,5-dioctyloxy-1,4-phenylenevinylene) as a macroinitiator for the nitroxyl-mediated controlled radical polymerization of flexible poly(styrene-*stat*-chlormethylstyrene) block.[2] The latter block was subsequently functionalized with C_{60} through atom-transfer radical addition:

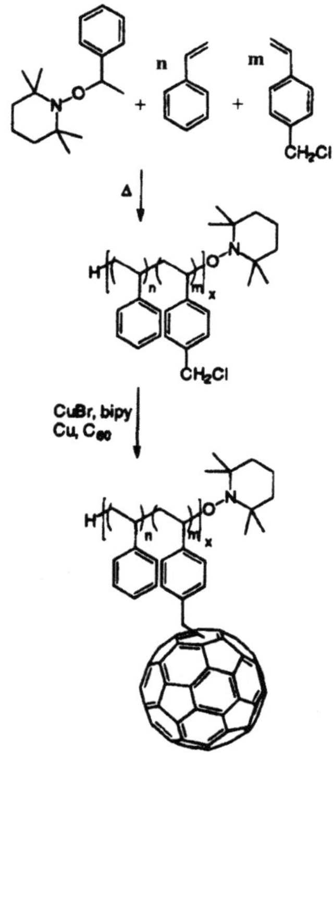

Series of statistical copolymers of styrene and 4-chloromethylstyrene (CMS) were synthesized, thus diluting the reactive groups in a random copolymer which still yields soluble functionalized materials.[2]

Random copolymers with different ratios of styrene to were synthesized using NMRP with TEMPO according to the following scheme:

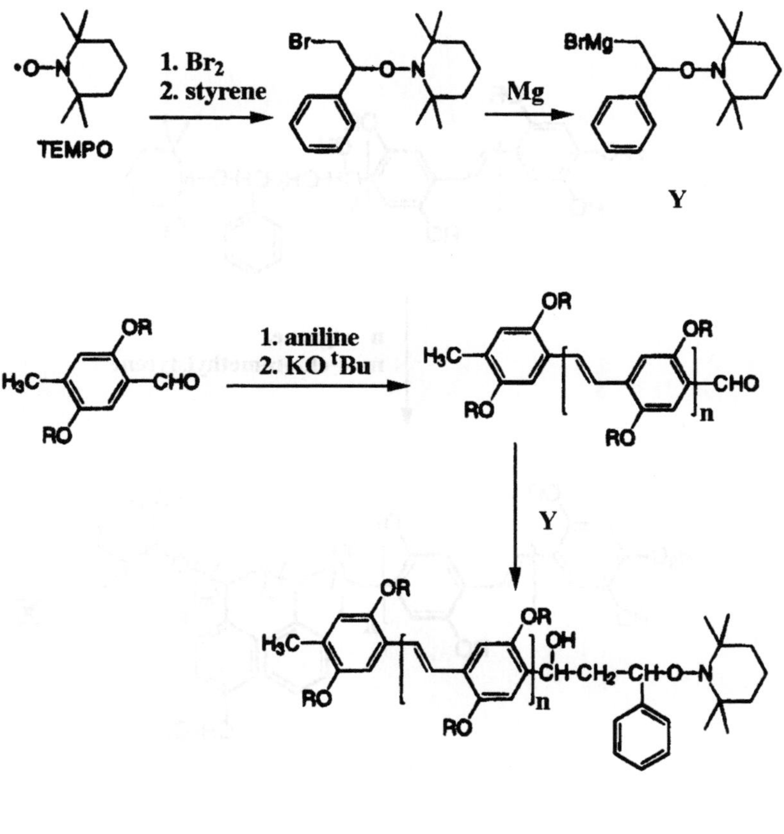

Molecular weights amounted to between 7 × 10^3 and 11 × 10^3 g mol^{-1} with polydispersities ranging from 1.12 to 1.22. These polymers were subsequently functionalized with C_{60}.

With increasing amounts of chloromethyl groups, the average distance between these groups decreases, and, hence, the possibility of multiple intramolecular additions on C_{60} will rise.

Scheme of synthesis of the alkoxyamine initiator, the aldehyde-functionalized PPV block, and their combination into the macroinitiator is as follows:[2]

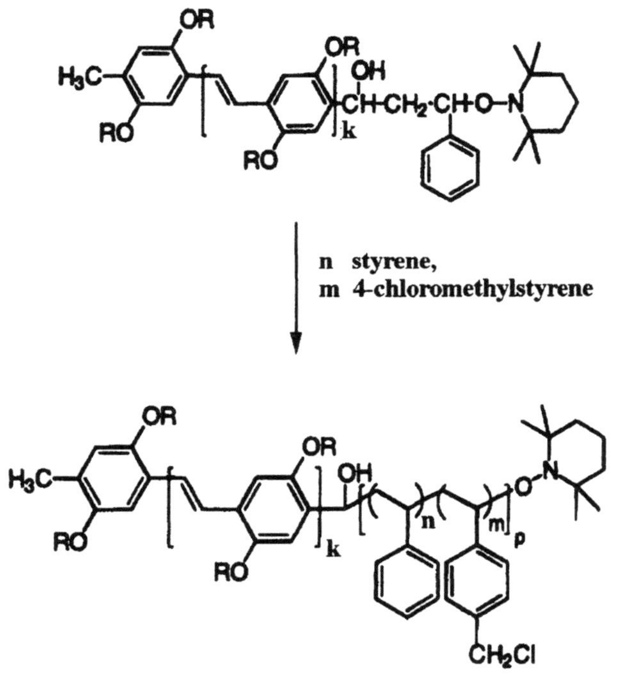

X

Combining the results of the two previous methods, the synthesis of random copolymer of styrene and CMS from PPV-based macroinitiator was carried out.

Subsequently, this block copolymer was functionalized with C_{60}:

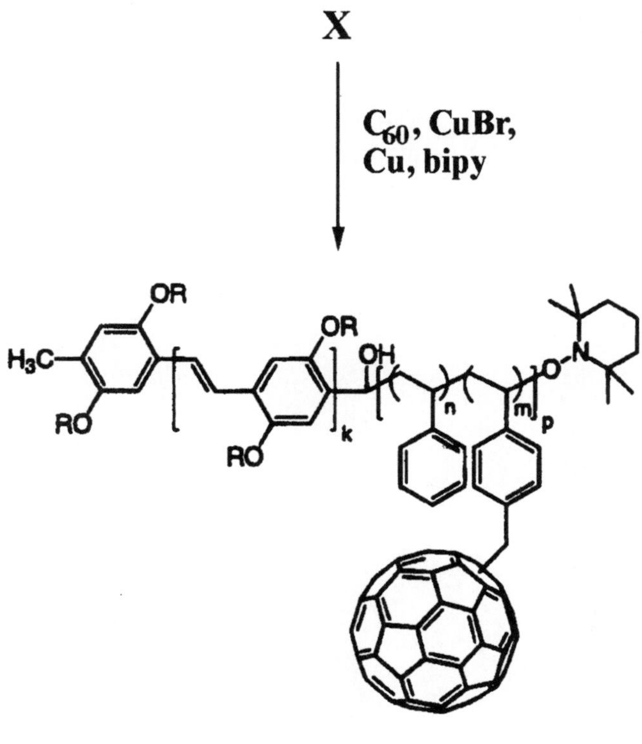

References

1. K. E. Geckeler, S. Samal, *Polym. Int.*, 1999, **48**, 743.
2. U. Stalmach, B. de Boer, C. Videlot, P. F. van Hutten, G. Hadziioannou, *J. Am., Chem. Soc.*, 2000, **122**, 5464.
3. T. Cao, S. E. Webber, *Macromolecules*, 1995, 28, 3741.
4. M. Seno, H. Fukunaga, T. Sato, *J. Pol. Sci.: Part A*, 1998, **36**, 2905.
5. W. T. Ford, T. Nishioka, S. C. McCleskey, T. H. Mourey, P. Kahol, *Macromolecules*, 2000, **33**, 2413.
6. H. Okamura, T. Terauchi, M. Minoda, T. Fukuda, K. Komatsu, , *Macromolecules*, 1997, **30**, 5279.
7. P. Zhou, G.-Q. Chen, H. Hong, F.-S. Du, Z.-C. Li, F.-M. Li, *Macromolecules*, 2000, **33**, 1948.
8. W. T. Ford, T. D. Graham, T. H. Mourey, *Macromolecules*, 1997, **30**, 6422.

9. C. E. Bunker, G. E. Lawson, Y.-P. Sun, *Macromolecules*, 1995, **28**, 3744.
10. B. J. Qu, G. Hawthorn, A. W. H. Mau, L. Dai, *J. Phys. Chem. B*, 2001, **105**, 2129.

Chapter 8

Ion Radical Reactions of Fullerenes

The description of the radical reactions of fullerenes would be not full without data on the radical reactions induced by one-electron transfer. In these reactions, both donor-acceptor properties of fullerenes and their ability to add free radicals are displayed.

Some examples of such reactions have already been under our consideration, e.g. electron transfer from $C_{60}^{-\bullet}$ on the proton followed by addition of H to C_{60}.[1]

C_{60} anions, such as $C_{60}^{-\bullet}$ or C_{60}^{2-} can be considered as potential nucleophiles since they react with electrophiles such as alkyl halides[2,3] and also act as electron donors in redox reactions.[4]

Formation of C_{60} adducts with two different alkyl groups via combination of electron transfer and S_N2 reactions was described by S. Fukuzumi et al. (Fig. 1).[5]

Oxidation of RC_{60}^{-} with I_2 leads to the formation of the dimers of the fullerenyl radical $RC_{60}^{\bullet}$ (see Chapter 2). These examples demonstrate an important place of C_{60} anions in radical reactions induced by one-electron transfer.

Besides, it was found that fullerene cation radicals formed under photochemical oxidation of fullerene are electrophile species and can abstract hydrogen atoms from different organic donors and attach the radicals formed. Reactions with the participation of fullerene cation radicals have a significant preparative value.

In this chapter we will describe the basic data on the EPR features of ion radicals of fullerenes. Full data can be found in reviews.[6,7]

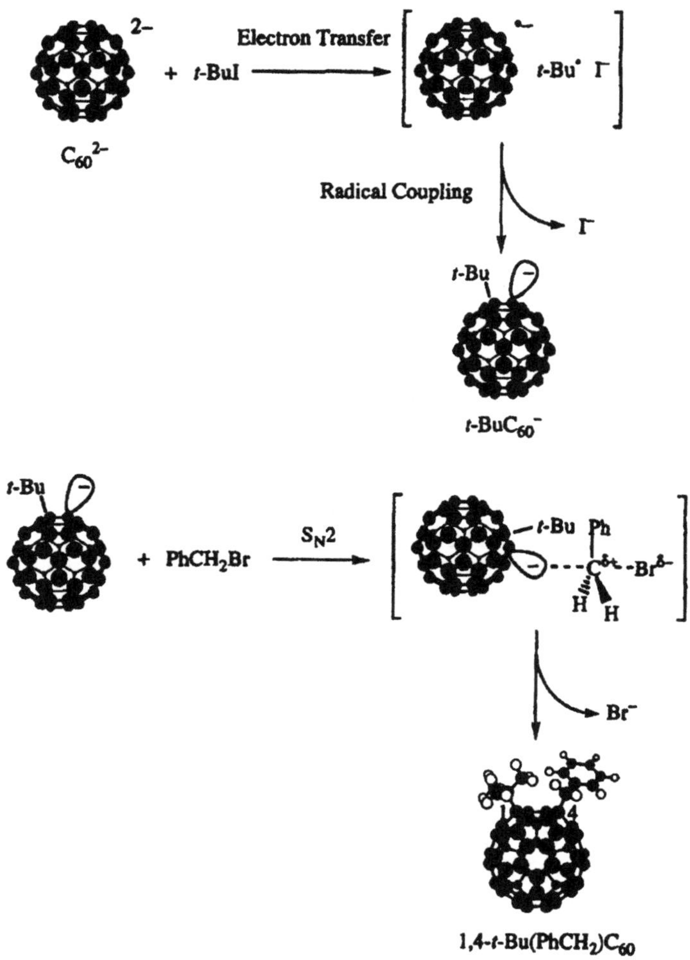

Figure 1. Formation of C_{60} adducts with two different alkyl groups via combination of electron transfer and S_N2 reactions.[5]

1.1 The EPR Spectra of C_{60} Anion Radicals

There are three features of the EPR spectrum of the $C_{60}^{-\bullet}$ anion radical:

1. A low g-factor;
2. An unusually broad signal at room temperature whose line width decreases markedly with decreasing temperature down to 77K;
3. Observable anisotropy at low temperature (4K).[6]

The most definitive EPR spectrum of the $C_{60}^{-\bullet}$ anion radicals (Fig. 2) was made on a well defined, air stable salt, $[(PPh_4)_2Cl]^+[C_{60}]^{-\bullet}$.[8]

The g-factor for this compound is 1.9991, significantly lower than that observed for a free electron (2.0023) and the values observed for most organic anion radicals. The unusually low value is attributed to spin-orbit coupling effects from unquenched angular momentum in Jahn–Teller-distored states.

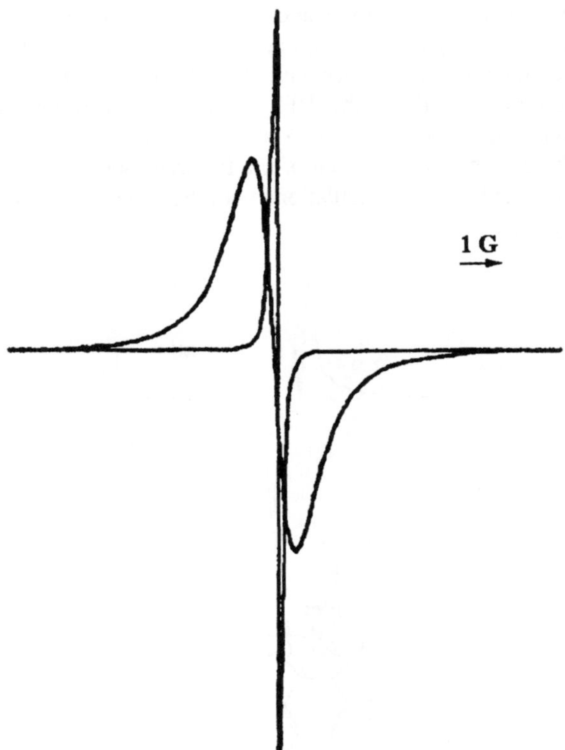

Figure 2. Solid-state EPR spectrum of C_{60}^- in $[PPh_4]_2Cl^+[C_{60}]^{-\bullet}$ at 300 K (broad line) and 80 K (narrow line). Adapted from ref 8.

The room-temperature signal was broad, with a peak-to-peak line width of 45 G decreasing to 7 G at 80 K. The dynamic nature of the small Jahn-Teller-distortion makes a main contribution to the observed effect.[8]

Another spectral pattern is observed for anion radicals derived from fullerene. Since the broad signals (at room temperature) are associated with the high symmetry of C_{60} and the $(t_{1u})^1$ degeneracy of $C_{60}^{-\bullet}$ and its Jahn-Teller distortion, a chemical modification of C_{60} that destroys this degeneracy is an origin of narrow signals at room temperature. In the EPR studies on the fullerene derivatives of the type $R_2C_{60}^{-\bullet}$, signals ascribable to such species have been reported to have g-factors close to 2.000 and line widths of 1 – 2 G.[9,10]

The radical anions of mono- and *bis-(trans*-1-fulleropyrrolidines) have been studied by continuos wave and pulsed X-band EPR and by high-frequency EPR. The hyperfine coupling constants of the ^{14}N nuclei and of the ^{13}C nuclei in natural abundance have been determined (Fig. 3).[11]

Calculation performed by the DFT method gives values for the ^{14}N hyperfine coupling very close to the experimental ones. The spin distribution is found to be on the equatorial belt of the fullerene sphere, as evidenced by comparing the results of the calculations with the experimentally determined ^{13}C *hfc*'s (Fig. 4).[11]

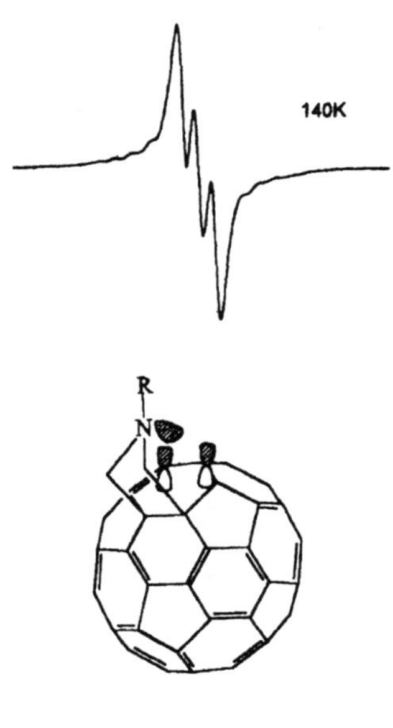

Figure 3. EPR spectra of the radical-anion of mono-fulleropyrrolidine. Adapted from ref 11.

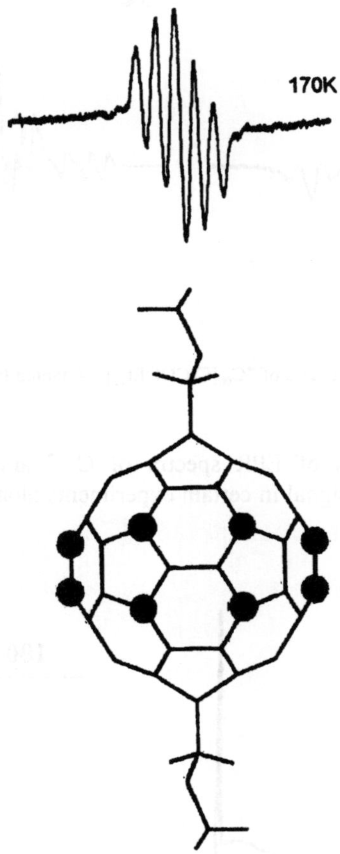

Figure 4. EPR spectra of the radical anion of *bis*-(*trans*-1-fulleropyrrolidine). Black balls mark atoms with the highest spin densities. Adapted from ref 11.

Bifunctional fullerene anion radical $^{-\bullet}C_{60}[C(COOEt)_2]_2$ exhibits much narrower lines.[9] Hyperfine coupling with ^{13}C nuclei in anion radicals $^{\bullet-}C_{60}[C(COOEt)_2]$ and $^{-\bullet}C_{60}[C(COOEt)_2]_2$ also indicate a considerable deviation of unpaired electron density distribution in the polyhedral fragments of these systems from spherical symmetry observed for the anion radical of C_{60} (Fig. 5).

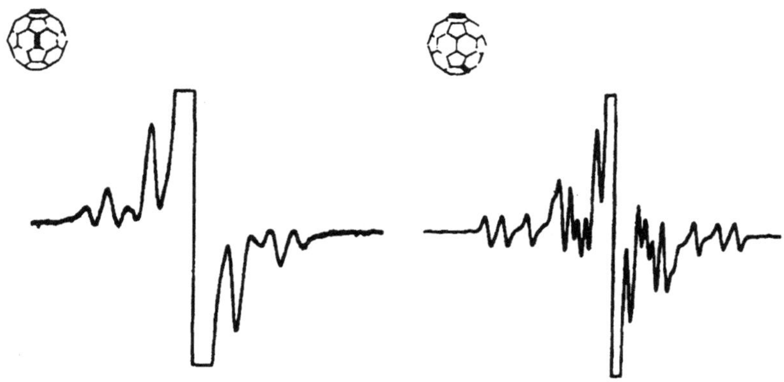

Figure 5. EPR spectra of $^{-\bullet}C_{60}[C(COOEt)_2]_2$. Adapted from ref 9.

One of the features of EPR spectra of $C_{60}^{-\bullet}$ anion radicals is the emergence of a narrow signal in certain experiments along with a broad line from $C_{60}^{-\bullet}$ (Fig. 6).[12]

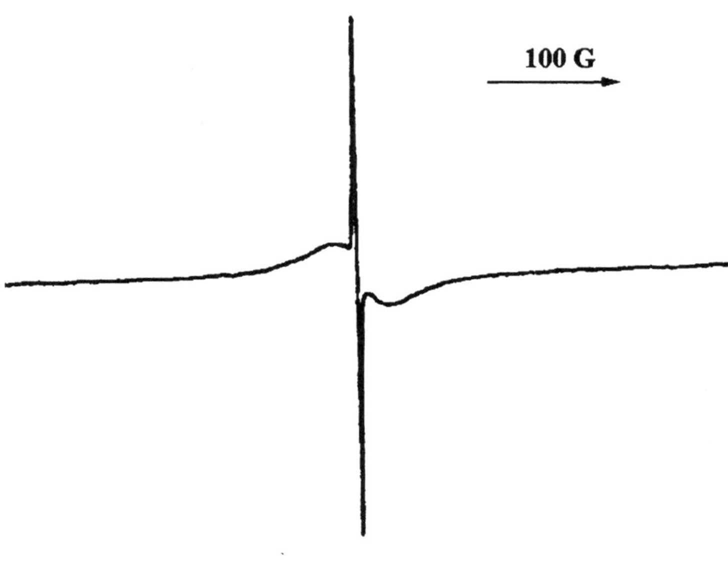

Figure 6. EPR spectrum of $C_{60}^{-\bullet}$ containing narrow signal. Adapted from ref 12.

Independent experiments showed that this line arises upon reduction of $C_{120}O$ as mixture in C_{60}.[6]

The value of |D| for a triplet dianion $C_{120}O^{2-}$ was found to be 1.3 mT. An average distance between two unpaired electrons in $C_{120}O^{2-}$ (13 Å) is comparable with the molecule size (ca. 16 Å).[13]

Spin density measurements by EPR indicate that the C_{60}^{2-} ion has a diamagnetic ground state.[14]

The high-temperature triplet is the only signal that can be ascribed to the C_{60}^{2-} ion (Fig. 7).[15,16]

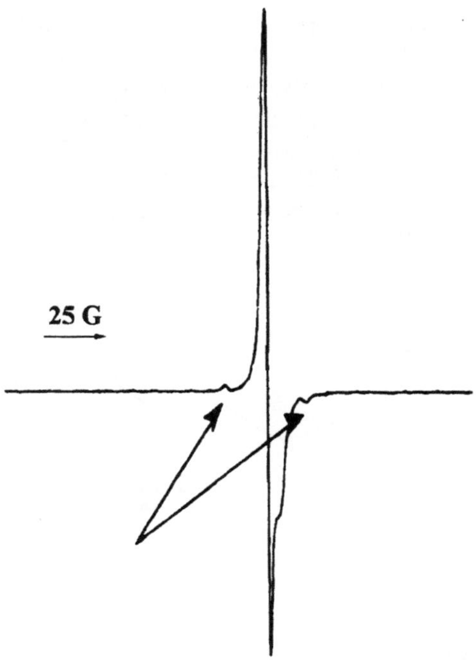

Figure 7. EPR spectrum of high temperature triplet of C_{60}^{2-}. Adapted from ref 15.

It was shown in the work[16] that this signal is due to the thermal occupation of an S = 1 state about 600 cm^{-1} above the ground state.

The similarities between the EPR spectra of anion-radicals $C_{60}^{-\bullet}$ and $C_{60}^{3-\bullet}$ are strong. The signals differ only in *g*-factor (for example 2.0025 for $C_{60}^{3-\bullet}$ and 1.9963 for $C_{60}^{-\bullet}$ in DMSO)[15] and in the details of the temperature dependence of the broad signal line width. The broad signal (room temperature) is the major component and is ascribed to anion-radical $C_{60}^{3-\bullet}$, having an S = 1/2 ground state.[17,18]

The EPR spectrum of $C_{70}^{-\bullet}$ [19] has remarkable line width features. In a well characterized $[PPh_4]_2C_{70}^{-\bullet}$ ΔH_{pp} is ca. 600 G at room temperature, reducing by orders of magnitude to ca. 2.G at liquid helium temperatures.[20]

The EPR spectrum $C_{76}^{-\bullet}$ has a relatively narrow signal (0.5 - 3.0 G) with a $g = 2.0023$[21] . This is consistent with the low symmetry of this fulleride.

It were also studied the EPR spectra of anion radicals derived from C_{78}[21] and C_{84}[22].

1.2 The EPR Spectra of Fullerene Cation Radicals

There have been many reports of EPR signals ascribable to $C_{60}^{+\bullet}$.[6] Most spectra arise from chemical or photochemical treatment of C_{60} under conditions where strongly oxidative chemistry is reasonably proposed.

It was shown in the work[23] that dissolution of C_{60} in concentrated sulfuric acid led to EPR spectra of paramagnetic species identified as dimers ($C_{120}^{+\bullet}$) or oligomers $(nC_{60})^{+\bullet}$. The EPR spectrum of these cation radicals at 77 K exhibited triaxial anisotropy of the g-factor, characterized by $g_1 = 2.0041$, $g_2 = 2.0019$, $g_3 = 2.0001$ (Fig. 8).

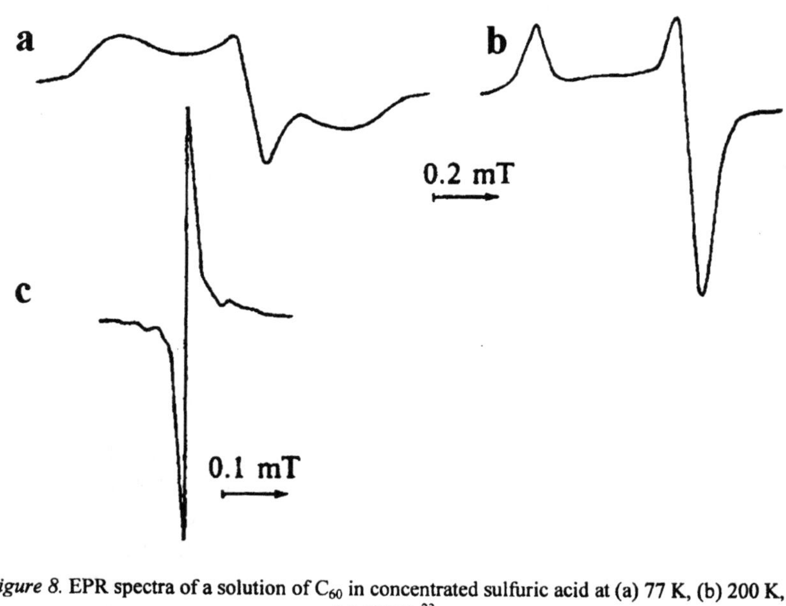

Figure 8. EPR spectra of a solution of C_{60} in concentrated sulfuric acid at (a) 77 K, (b) 200 K, (c) 330 K.[23]

As the temperature increases, the frequency of rotation of a paramagnetic species increases, and the line is broadened. Thus, spectrum (*a*) transforms into an asymmetric line. At temperatures above 200 K, the latter transforms into spectrum (*b*) typical of a system with axial symmetry of the *g*-factor.

At 330K, the line width is 0.0175 mT, and satellite signals are recorded at the edges of the central line. They are components of the hyperfine structure caused by coupling with the ^{13}C nuclei. The splitting values are a_l = 0.175 mT, a_2 = 0.100 mT, and a_3 = 0.030 mT, and g = 2.0026.[23]

The frequency of rotation of the dimer in solid matrix around the X axis should be much higher than that around the Y and Z axes:

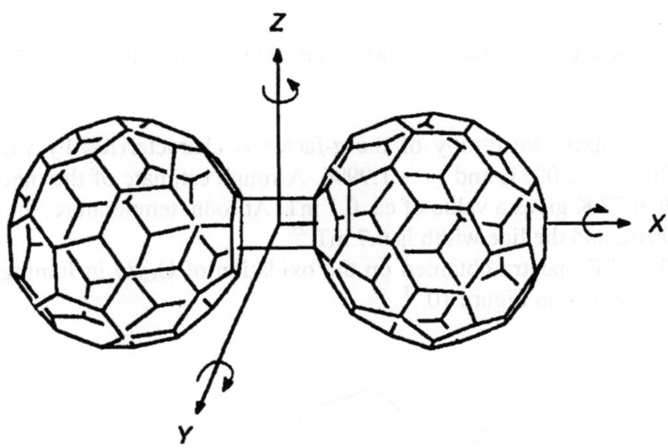

This rotation should average the positions of the EPR lines corresponding to particular orientations of the Y and Z axes in the magnetic field. Hence, the arrangement of the X in parallel to the vector of the magnetic field intensity corresponds to the signal $g_1 = g_{||}$, while the averaging of the position of the g_2 and g_3 signals gives $g_{\perp}$. EPR spectra obtained on C_{120} oxidation in fuming sulfuric acid are coincide completely with the pattern observed on the oxidation of C_{60}.

The EPR spectrum of the sample obtained by the reaction of toluene solution of C_{60} with sulfuric acid were assigned to the radical cation $C_{60}^{+\bullet}$. The EPR spectrum of the sample obtained by this method recorded at 77 K is shown in Figure 9.[23]

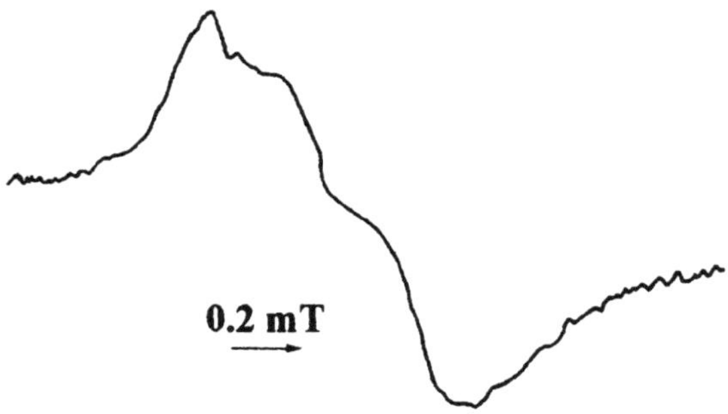

Figure 9. EPR spectrum of C_{60} in a mixture of toluene and sulfuric acid at 77 K.[23]

The triaxial anisotropy of the g-factor is characterized by values g_1 = 2.0036, g_2 = 2.0021, and g_3 = 1.9996. A rough estimate of the inherent line width at 77 K gives a value of ca. 0.1 mT. At room temperature, the g-factor is 2.0016, and the line width is 0.7 mT.[23]

The EPR spectra obtained on the oxidation of $C_{120}O$ in fuming sulfuric acid are shown in Figure 10.[24]

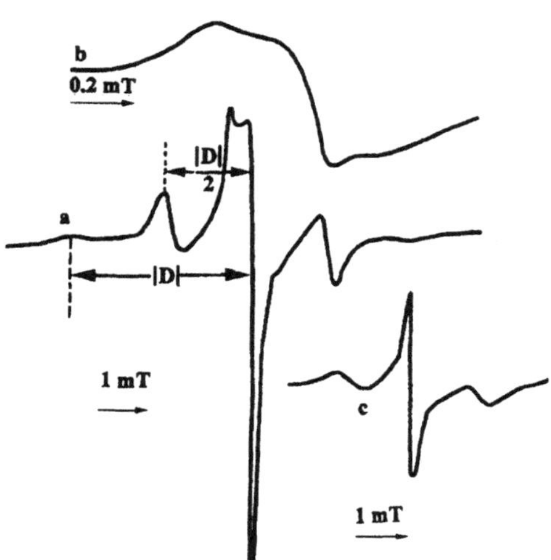

Figure 10. EPR spectra of $C_{120}O^+/C_{120}O^{2+}$ generated by the oxidation of $C_{120}O$ in fuming sulfuric acid at (a) 77 K, (c) 300 K; (b) selected spectrum of $C_{120}O^+$ at 77 K.[24]

Their appearance suggests that two species having spins $S = 1/2$ and $S = 1$ are present in solution. In the spectrum at 77 K (Fig. 10 a,b), a central anisotropic line ($g_1 = 2.0038$, $g_2 = 2.0020$, and $g_3 = 2.0002$) corresponds to a cation-radical, $C_{120}O^{+\bullet}$.

Four sidelines in Fig. 10a are characteristic of the spectrum of randomly oriented species with $S = 1$. This spectrum assigns them to a triplet dication-radical $C_{120}O^{2+\bullet}$ with two unpaired electrons with parallel spins, the first electron being situated on the one cage and the second on the other cage.

The zero-field parameters $|D|$ and $|D/2|$ determine the positions of these sidelines corresponding to the parallel and perpendicular orientations of $C_{120}O^{2+\bullet}$ with regard to a direction of the external magnetic field. The value of $|D|$ (2.95 mT) indicates substantial interaction between the two electrons.

The appearance of the sidelines suggests also the axial symmetry in the spin density distribution in each fullerenyl fragment of $C_{120}^{2+\bullet}$. It is seen in Fig. 10c that the rotation of $C_{120}O^{2+\bullet}$ at room temperature does not average the dipole-dipole interaction to zero.[24]

Hence, the speed of tumbling rotations of the dication at room temperature is less than $|D|$ or, if expressed in frequency units, less than $8.4 \times 10^7\ s^{-1}$.

It is interesting to note that the value of $|D|$ for a triplet dianion-radical (1.3 mT)[12] is far less than for a triplet dication-radical. The difference in $|D|$ implies a significant distinction between the spin density distributions in $C_{120}O^{2-\bullet}$ and $C_{120}O^{2+\bullet}$.[24]

Cation radical *tris*(2,4-dibromophenyl)amine was used as oxidant for the synthesis of fullerenium cation-radical .[6] A suitable anion was the carborane $CB_{11}H_9Br_6^-$, an inert anion with low nucleophilicity:

$$C_{76} + [Ar_3N^{+\bullet}][CB_{11}H_6Br_6^-] \rightarrow [C_{76}^{+\bullet}][CB_{11}H_6Br_6^-] + Ar_3N$$

Strong oxidant such as cation-radical of a hexabrominated phenylcarbazole was used for the oxidation of C_{60} to cation-radical. $[C_{60}^{+\bullet}][CB_{11}H_6Cl_6^-]$ has been characterized by NIR ($\lambda_{max} = 980$ nm), and EPR.[6]

Cation radical $C_{60}^{+\bullet}$ can act as intermediate in reactions with electron transfer. For the first time, new and unexpected chemistry of cation radical $C_{60}^{+\bullet}$ in solution was described in the work.[25].

As typical procedure, a mixture of C_{60}, H donor reactant, and photoelectron sensitizer were irradiated with laser at 425 nm in a suitable solvent.

Comparative studies involving well-known photoinduced electron transfer (PET) sensitizers such as, 1.4-dicyanonaphthalene (DCN), *N*-methylacridinium hexafluorophosphate (NMA^+) and 2,4,6-triphenylpyrylium tetrafluoroborate (TPP^+) revealed the suitability of all sensitizers in the presence of biphenyl as a cosensitizer with increasing velocity.[26]

Both cationic sensitizers (by using an electron shift) and cosensitizers lead to an increased separation of $C_{60}{}^{+\bullet}$ radical cation and the reduced sensitizer and therefore decrease the probability of back electron transfer.

The proposed mechanizm for the photosensitized addition of the addends is outlined in Scheme:[25]

$$\mathrm{DCA} \xrightarrow{h\nu} \mathrm{DCA^*}$$

$$\mathrm{DCA^*} + \mathrm{C_{60}} \longrightarrow [\mathrm{DCA^*}\bullet\bullet\bullet\bullet\mathrm{C_{60}}]$$

$$[\mathrm{DCA^*}\bullet\bullet\bullet\bullet\mathrm{C_{60}^*}] \longrightarrow [\mathrm{C_{60}}]^{\bullet+}\bullet\bullet\bullet\bullet[\mathrm{DCA}]^{-\bullet}$$

$$[\mathrm{C_{60}}]^{\bullet+}\bullet\bullet\bullet\bullet[\mathrm{DCA}]^{-\bullet} \xrightarrow{\mathrm{CH_3CN}} [\mathrm{C_{60}}]^{\bullet+}_{\mathrm{solv}} + [\mathrm{DCA}]^{-\bullet}_{\mathrm{solv}}$$

$$[\mathrm{C_{60}}]^{\bullet+}_{\mathrm{solv}} + \mathrm{RH} \longrightarrow [\mathrm{HC_{60}}]^{+}_{\mathrm{solv}} + \mathrm{R^\bullet}$$

$$[\mathrm{HC_{60}}]^{+}_{\mathrm{solv}} + [\mathrm{DCA}]^{-\bullet}_{\mathrm{solv}} \longrightarrow [\mathrm{HC_{60}}]^{\bullet}_{\mathrm{solv}} + [\mathrm{DCA}]_{\mathrm{solv}}$$

$$\mathrm{HC_{60}}^\bullet + \mathrm{R^\bullet} \longrightarrow \mathrm{HC_{60}R}$$

The formation of fullerene adducts of H-donors such as aldehyde, ethers, alcohols, toluene, cyclohexane, cyclohexene, *N,N*-dimethylformamide, 1,3-dioxolane, phenylacetaldehyde, metyl formate, *tert*-butanol, propionic acid, glycol when C_{60} and sensitizer are irradiated at 425 nm indicates that, under these conditions, the fullerene radical cation can behave as a typical free radical and abstract hydrogen from a suitable hydrogen donor.[26]

Let us discuss some features for reactions with $C_{60}^{+\bullet}$. As hydrogen abstraction from propionic acid usually takes place at the α-position, the cationic character of $C_{60}^{+\bullet}$ obviously changes reaction pathways and affords hydrogen abstraction at unusual positions.

The reaction of glycol follows a similar pathway:[26]

$$RCOOH \xrightarrow{-H} RCOO^{\bullet} \xrightarrow{-CO_2} R^{\bullet}$$

$$HO\text{-}CH_2\text{-}CH_2\text{-}OH \xrightarrow{-H} HO\text{-}CH_2\text{-}CH_2\text{-}O^{\bullet} \xrightarrow{-CH_2O} HO\text{-}CH_2^{\bullet}$$

Similarly the loss of an hydrogen from the hydroxyl function of $(CH_3)_3COH$ results in the formation of $H_3C^{\bullet}$.

The reason for hydrogen abstraction from such an unusual position was described as follows:[26]

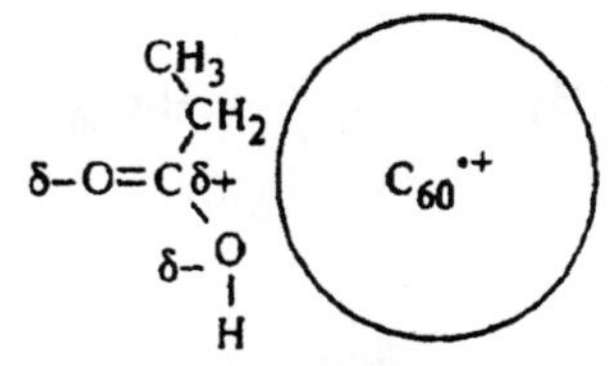

favored transition state

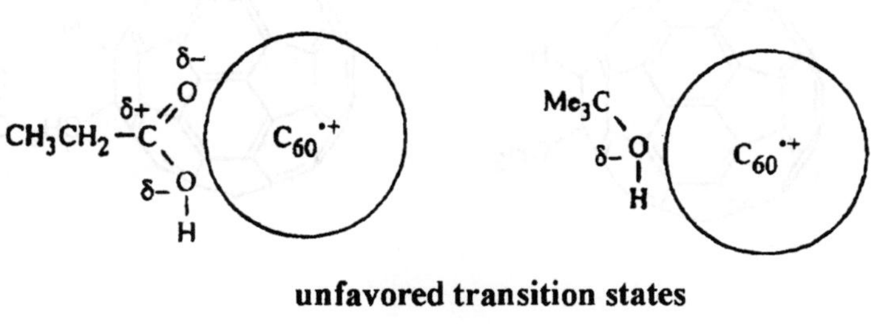

unfavored transition states

The bond energy of H–O (440 kJ mol^{-1}) is higher than that of the H–C (401 kJ mol^{-1}). Although the homolytic break of H-C bond should therefore be favoured, α-C-radical of propionic acid is not formed.

The positive charge of radical cation lowers the energy of transition state, leading to hydrogenabstraction from the hydroxyl position of both $(CH_3)_3COH$ and $HOCH_2CH_2OH$.[26]

Following this argumentation, the products of the reaction of *tert*-butanol and C_{60} under PET-conditions can be explained:[26]

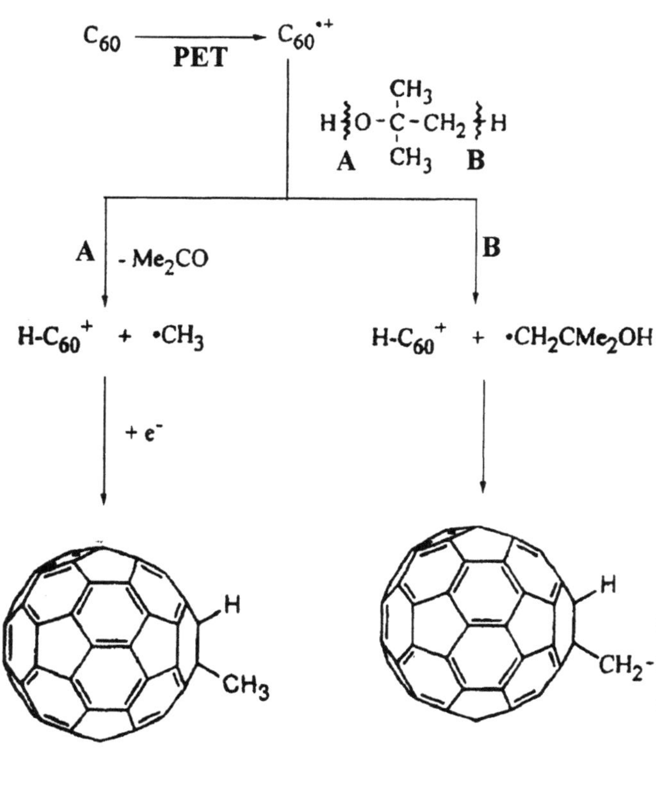

Homolytic bond breaking at the H–O position leads to the formation of $HC_{60}CH_3$, while the abstraction at the H–C position results in $HC_{60}CH_2(CH_3)_2OH$.

Both time-resolved laser flash photolysis studies and EPR spectroscopic investigations clearly demonstrate the formation of $C_{60}^{+\bullet}$ using PET.[27]

References

1. T. Tanaka, K. Komatsu, *J. Chem. Soc., Perkin Trans. 1*, 1999, 1671.
2. C. Caron, R. Subramanian, F. D'Souza, J. Kim, W. Kutner, M. T. Jones, K. M. Kadish, *J. Am. Chem. Soc.*, 1993, **115**, 8505.
3. R. Subramanian, F. D'Souza, M. N. Vijayashree, X. Gao, M. T. Jones, M. D. Miller, K. L. Krause, T. Suenobu, S. Fukuzumi, *J. Phys. Chem.*, 1996, **100**, 16327.
4. Y. Huang, D. D. M. Wayner, *J. Am. Chem. Soc.*, 1993, **115**, 367.
5. S. Fukuzumi, T. Suenobu, T. Hirasaka, R. Arakawa, K. M. Kadish, *J. Am. Chem. Soc.*, 1998, **120**, 9220.
6. C. A. Reed, R. D. Bolskar, *Chem. Rev.*, 2000, **100**, 1075.
7. S. S. Eaton, G. R. Eaton, *Appl. Magn. Res.*, 1996, **11**, 155.
8. P.-M. Allemand, G. Srdanov, A. Koch, K. Khemani, F. Wudl, *J. Am. Chem. Soc.*, 1991, **113**, 2780.
9. V. Brezova, A. Stasko, P. Rapta, D. M. Guldi, K.-D. Asmus, K.-P. Dinse, *Magn. Res. Chem.*, 1997, **35**, 795.
10. M. Baumgarten, L. Gherghel, *Appl. Magn. Reson.*, 1996, **11**, 171.
11. A. Zoleo, A. L. Maniero, M. Prato, M. G. Severin, L. C. Brunel, K. Kordatos, M. Brustolon, *J. Phys. Chem. A*, 2000, **104**, 9853.
12. M. A. Greaney, S. M. Gorun, *J. Phys. Chem.*, 1991, **95**, 7142.
13. A. L. Balch, D. A. Costa, W. R. Fawcett, K. Winkler, *J. Phys. Chem.*, 1996, **100**, 4823.
14. H. Moriyama, H. Kobayashi, A. Kobayashi, T. Watanabe, *J. Am. Chem. Soc.*, 1993, **115**, 1185.
15. P. D. W. Boyd, P. Bhyrappa, P. Paul, J. Stinchcombe, R. D. Bolskar, Y. Sun, C. A. Reed, *J. Am. Chem. Soc.*, 1995, **117**, 2907.
16. P. C. Trulove, R. T. Garlin, G. R. Eaton, S. S. Eaton, *J. Am. Chem. Soc.*, 1995, **117**, 6265.
17. P. Bhyrappa, P. Paul, J. Stinchcombe, P. D. W. Boyd, C. A. Reed, *J. Am. Chem. Soc.*, 1993, **115**, 11004.
18. M. Baumgarten, A. Gugel, L. Gherghel, *Adv. Mater.*, 1993, **5**, 458.
19. D. Dubois, K. M. Kadish, S. Flanagan, R. E. Haufler, L. P. F. Chibante, L. J. Wilson, *J. Am. Chem. Soc.*, 1991, **113**, 4364.
20. A. Penicaud, A. Perez-Benitez, R. Escudero, C. Coulon, *Solid State Commun.*, 1995, **96**, 147.
21. J. A. Azamar-Barrios, P. E. Munoz, A. Penicaud, *J. Chem. Soc., Faraday Trans.*, 1997, **93**, 3119.
22. P. L. Boulas, M. T. Jones, R. S. Ruoff, D. C. Lorentz, R. Malhorta, D. S. Tse, K. M. Kadish, *J. Phys. Chem.*, 1996, **100**, 7573.
23. S. P. Solodovnikov, *Russ. Chem. Bull.*, 1998, **47**, 2302.
24. S. P. Solodovnikov, B. L. Tumanskii, V. V. Bashilov, V. I. Sokolov, S. F. Lebedkin, W. Kratschmer, *Chem. Phys. Lett.*, 1999, **303**, 387.
25. G. Lem, D. I. Schuster, S. H. Courtney, Q. Lu, S. R. Wilson, *J. Am. Chem. Soc.*, 1995, **117**, 554.

26. C. Siedschlag, H. Luftmann, C. Wolff, J. Mattay, *Tetrahedron*, 1999, **55**, 7805.
27. C. Siedshlag, L. Dunsch, J. Mattay , *J. Informat. Record.*, 2000, **25**, 265.

Chapter 9

Azafullerenes in Radical Reactions

Azafullerenes are fullerenes in which one carbon atom is replaced by a nitrogen one (Fig. 1).[1,2] This leads to the formation of the azafullerenyl radical, $^{\bullet}C_nN$ (n = 59, 69), which is isoelectronic with the $C_n^{-\bullet}$.

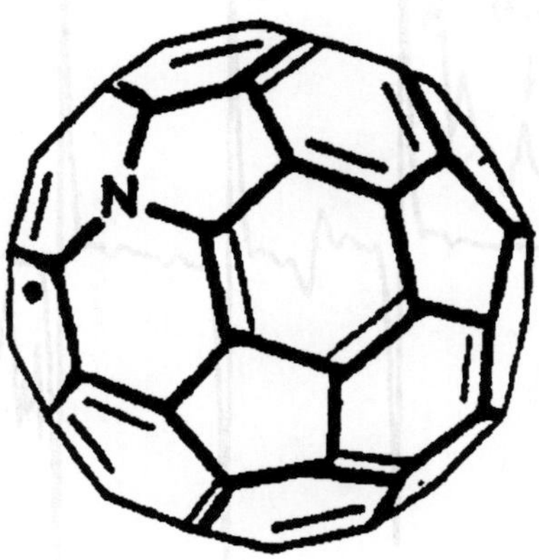

Figure 1. Azafullerene $^{\bullet}C_{59}N$.

Interest in these molecules arises from the fact that the substitution of a carbon with a trivalent nitrogen is analogous to doping of semiconductors with an impurity. The addition of an impurity to such a system is a strong perturbation and might give rise to the formation of states deep within the gap.[3] Thus, a route for variation of the electronic properties of fullerenes is possible.

$^{\bullet}C_{59}N$ is a very reactive radical, which is found to dimerize or to form $C_{59}NH$ in the presence of reducing agents. However, solid solutions of $^{\bullet}C_{59}N$ in C_{60} were produced, which allowed to measure both ^{14}N and ^{13}C isotropic hyperfine coupling constants (Fig. 2).[4]

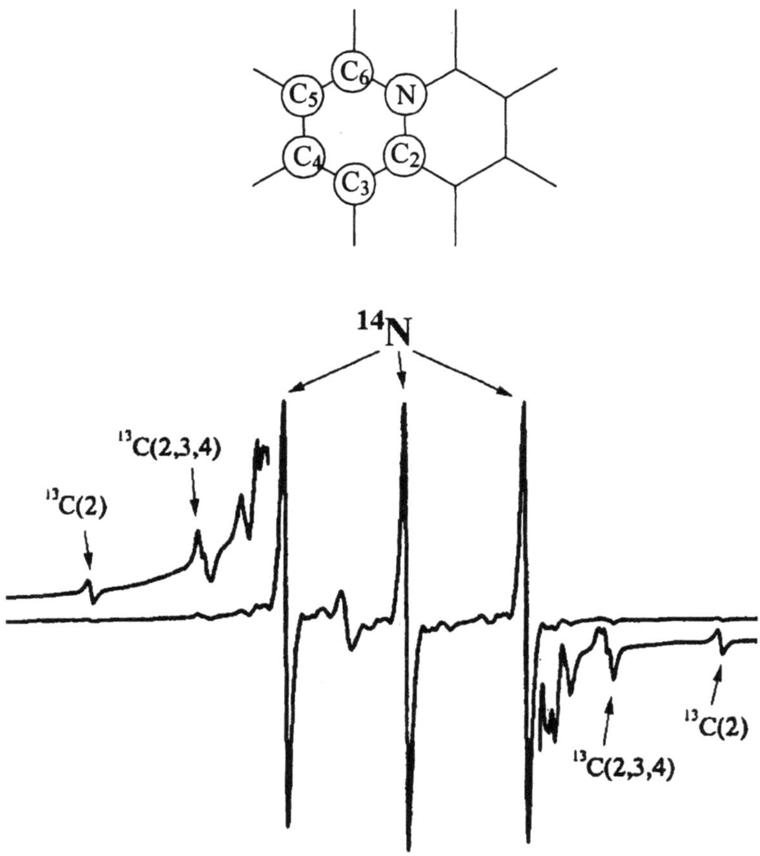

Figure 2. EPR spectrum of $^{\bullet}C_{59}N$ substituted in C_{60} at 290 K. Adapted from ref 4.

The intermolecular bond strength between the two molecular fragments in $(C_{59}N)_2$ (Fig. 3) is about 7 kcal mol^{-1}, that is in reasonable agreement with theoretical predictions.[3,5] This bond can break easily upon irradiation to produce the radicals $^{\bullet}C_{59}N$. Information about the local structure of the binding sites in these dimers can be obtained from the EPR spectra of the radicals formed.

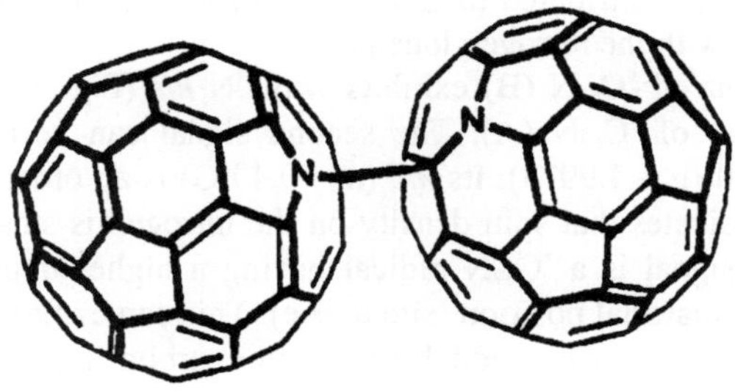

Figure 3. $(C_{59}N)_2$ dimer.

Upon light irradiation of the azafullerene dimers, the EPR spectra exhibit clear evidence of radical formation.[3] The three sharp lines are typical of a single ^{14}N hyperfine splitting with a hyperfine coupling constant of 3.73 G. The integrated intensity of the lines increases linearly with irradiation power, as expected for a monomolecular process. The signal disappears upon switching off the light and can be observed repeatedly without loss of intensity. g-Factor was found to be 2.0013 that is higher than the value reported for the C_{60} radical anion (see Chapter 3).[3]

As distinct from C_{60}, there are five non-equivalent carbon atoms in C_{70}, and five different $C_{69}N$ isomers can be obtained.[3] The formation of all five radicals can lead to 15 isomeric $(C_{69}N)_2$. Two isomers of $(C_{69}N)_2$ have been investigated. They differ in the position of the nitrogen atom: on the pole in *A* and one bond toward the equator in *B*:

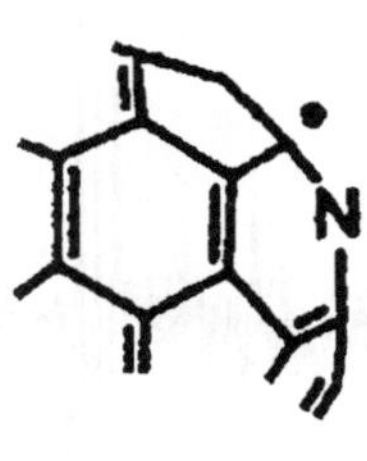

A

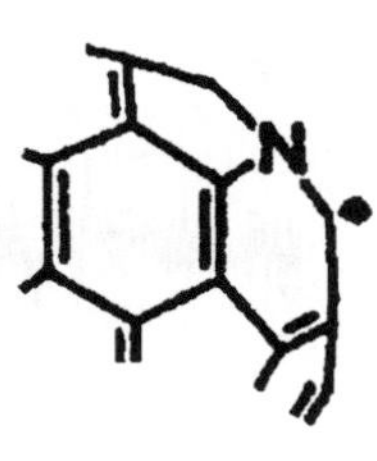

B

Photolysis of $(C_{69}N)_2$ (*A* isomer) leads to the EPR spectrum of isomer *A* (g = 2.0024, *a* = 4.74 G) (Fig. 4a).

The larger value of the hyperfine coupling constant for $^{\bullet}C_{69}N$, compared to that for $^{\bullet}C_{59}N$, is attributed to greater localization of the radical and thus greater overlap with the nitrogen lone pair.

The spectrum of $^{\bullet}C_{69}N$ (B) exhibits two ^{14}N *hfs* (Fig. 4b).[3] The first is identical to that of $^{\bullet}C_{69}N$ (A). The second signal can be attributed to a different radical (g = 1.9973). Its *hfc* (*a* = 0.49 G) is an order of magnitude smaller that indicates that spin density on the nitrogen is small. A possible origin of this signal is a $^{\bullet}C_{69}N$ radical having a higher unpaired electron density at the equatorial position (site d or e). This can explain the small hfc arising from a larger separation between the carbon and nitrogen atoms, leading to small overlap between their wave-functions. A more reasonable explanation is that the radical of isomer ***B*** has two different electron localization sites, where the unpaired electron in one of them is forced to be on the equatorial site because of structural strain in the molecule. The occurrence of two different electron localization sites in radical ***B*** and only one site in radical ***A*** might be attributable to difference in strain energy in different regions of the parent C_{70} molecule.[3]

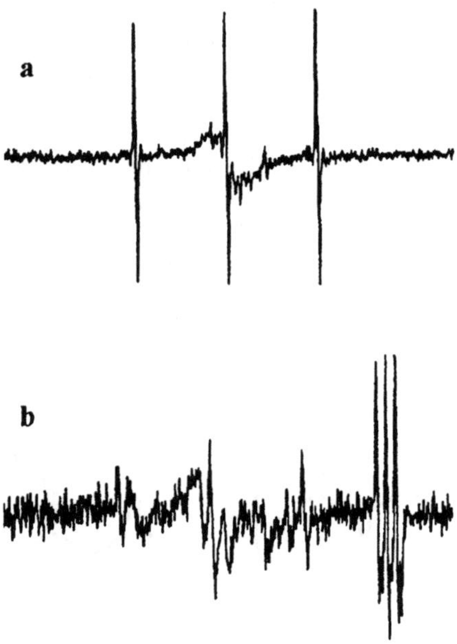

Figure 4. EPR spectra of $^{\bullet}C_{69}N^A$ (a) and $^{\bullet}C_{69}N^B$ (b). Adapted from ref 3.

There is only one radical reaction known for $^{\bullet}C_{59}N$. If the reaction is carried out in the absence of either oxygen or an acid and in the presence of diphenylmethane, which has a readily abstractable hydrogen, then radical coupling occurs between $^{\bullet}C_{59}N$ and $^{\bullet}Ph_2CH$ (Fig. 5):[6]

$$(C_{59}N)_2 \xrightarrow{\text{ODCB}, \Delta} 2\,C_{59}N^{\bullet}$$

$$C_{59}N^{\bullet} + Ph_2CH_2 \longrightarrow HC_{59}N + Ph_2CH^{\bullet}$$

$$Ph_2CH^{\bullet} + C_{59}N^{\bullet} \longrightarrow (Ph_2C)C_{59}N$$

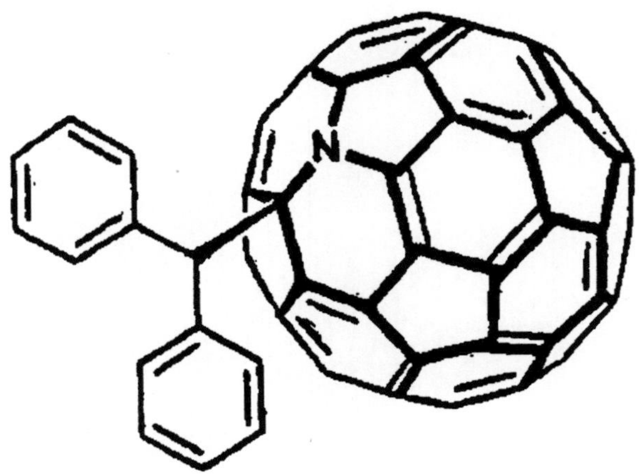

Figure 5. Structure of $(Ph_2C)C_{59}N$.

This reaction also has a potential for the formation of a wide range of interesting derivatives.

References

1. R. Taylor, *Lecture Notes on Fullerene Chemistry: A Handbook for Chemists*, Imperial College Press, London, 1999.
2. U. Reuther, A. Hirsch, *Carbon*, 2000, **38**, 1539.
3. K. Hasharoni, C. Bellavia-Lund, M. Keshavarz-K., G. Srdanov, F. Wudl, *J. Am. Chem. Soc.*, 1997, **119**, 11128.

4. F. Fulop, A. Rockenbauer, F. Simon, S. Pekker, L. Korecz, S. Garaj, A. Janossy, *Chem. Phys. Lett.*, 2001, **334**, 233.
5. F. Simon, D. Arcon, N. Tagmatarchis, S. Garaj, L. Forro, K. Prassides, *J. Phys. Chem. A*, 1999, **103**, 6969.
6. C. Bellavia-Lund, R. Gonzalez, J. C. Hummelen, R. G. Hicks, A. Sastre, F. Wudl, *J. Am. Chem. Soc.*, 1997, **119**, 2946.

Chapter 10

Endohedral Metallofullerenes in Radical and Ion Radical Reactions

Endohedral metallofullerenes containing metal atoms encapsulated in fullerene cages, in particular $M@C_{82}$ with M = Sc, Y, La, have been intensely studied in the last years (Fig. 1).[1-5] Theoretical and experimental results, in particular those obtained by EPR spectroscopy, have shown that these metallofullerenes can be described as $M^{3+}@C_{82}^{3-}$ ionic structures with the metal ion being located off-center and electrostatic interactions playing the main role in metal-to-cage bonding.[1-5]

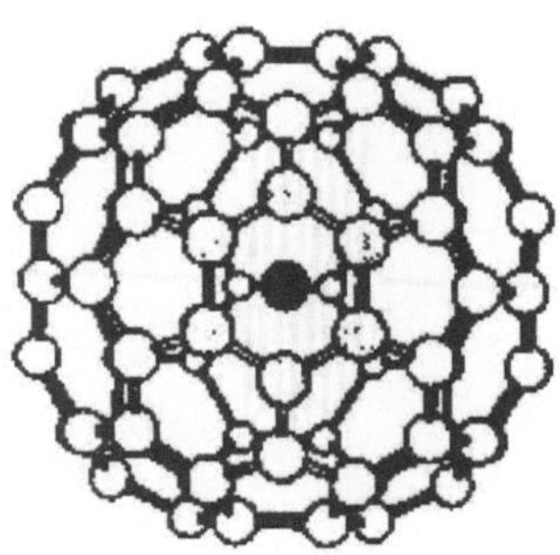

Figure 1. $M@C_{82}$ (M = La, Y).

The functionalization of metallofullerenes is of interest to increase their solubility and for their application as new types of organic ferromagnets, nonliner optical materials, etc. Let us consider briefly several examples of reactivity of $M@C_{82}$, where M = Y, La.

1.1 Addition Reactions

Addition of phosphoryl radicals to $La@C_{82}$ was studied:[6]

$$M^{3+}@C_{82}^{3-\bullet} + {}^{\bullet}P(O)(OPr^i)_2 \longrightarrow M^{3+}@C_{82}^{3-\bullet\bullet}\text{-}P(O)(OPr^i)_2$$

The ground state of resultant adduct seems to be singlet and therefore the addition of one or odd number of radicals to metallofullerenes makes them diamagnetic. The addition of even number of free radicals leads to paramagnetic products.

A model addition of two radicals may be exemplified by cycloaddition of silicon heterocycles to $La^{3+}@C_{82}^{3}$:[7]

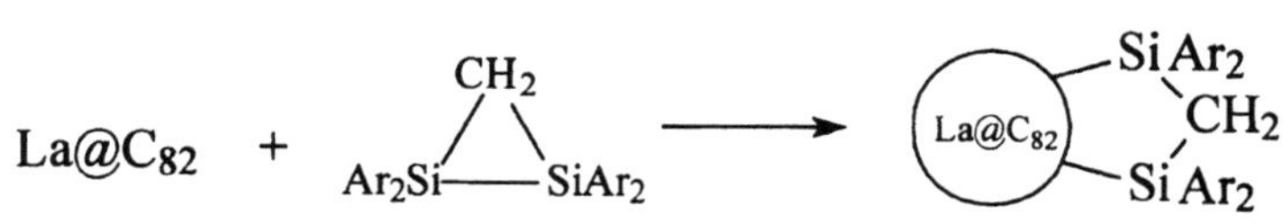

Fullerene C_{82} comprises a set of nonequivalent double bonds and the cycloaddition results in several isomers. EPR study has shown the reaction products to be paramagnetic, with *hfc* constants with La nucleus and *g*-factors being different for all isomers. Therefore the EPR spectrum of reaction product is a complex superposition of signals from different isomers (Fig. 2).

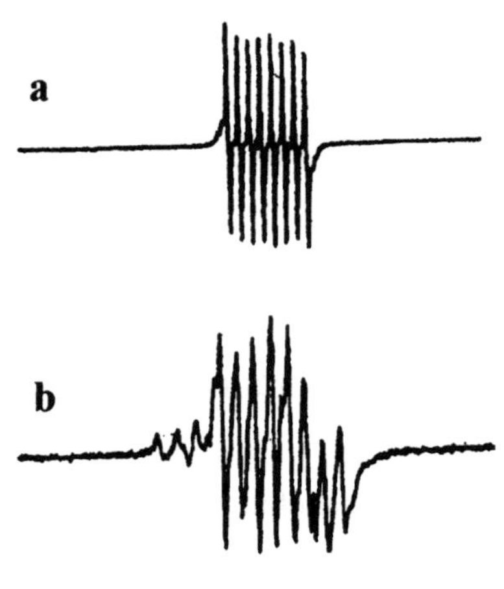

Figure 2. EPR spectra of (a) $La@C_{82}$ and (b) reaction mixture. Adapted from ref 7.

Similar pattern was observed for methanofullerene derivatives of $La@C_{82}$:[8]

$$La@C_{82} + Ph_2CN_2 \longrightarrow La@C_{82}CPh_2 + N_2$$

The EPR spectrum of reaction mixture is a superposition of four octets (Fig. 3), whose parameters are given in Table 1.

Figure 3. EPR spectrum of the reaction mixture containing $La@C_{82}$ and diphenyldiazomethane. Adapted from ref 8.

Table 1. EPR parameters for octets from $La@C_{82}$ and adducts observed.[8]

Octet	*Hfc* constant, G	g-factor
$La@C_{82}$	1.15	2.0012
A	1.18	2.0006
B	1.17	2.0009
C	1.14	2.0015
D	1.13	2.0008

Multiple addition of phenyl radicals to $La@C_{82}$ produces a set of products $La@C_{82}(Ph)_n$ (n = 1,2,3,4...):[9]

$$Ph(CO)O\text{-}O(CO)Ph \xrightarrow{410K} {}^{\bullet}Ph + CO_2$$

$$La@C_{82} + n\,({}^{\bullet}Ph) \longrightarrow La@C_{82}Ph_n$$

The superposition of EPR spectra from compounds with different number of addends (n = 2, 4, ...) and a large number of possible isomers are the reasons of a broadened singlet (g = 2.0026, $H_{pp} \sim 4.0$ G), without *hfi* with La nucleus.

1.2 Oxidation and Reduction Reactions

Oxidation of M@C_{82} with concentrated sulfuric acids has been studied.[10] EPR spectra of paramagnetic cations $[M@C_{82}]^{2+}$ have been recorded (Fig. 4).

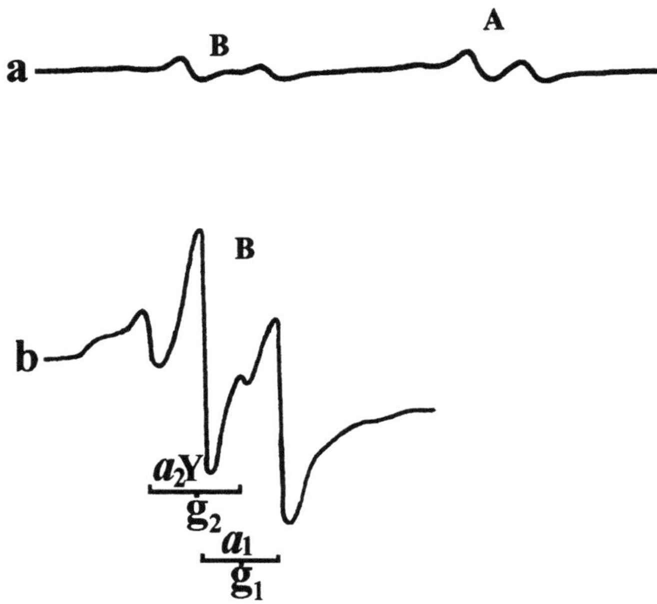

Figure 4. EPR spectra of (a) Y@C_{82} (A) and $Y^{3+}@C_{82}^-$ (B) obtained by oxidation with oleum in the beginning of reaction, (b) in 2 min (2 isomers of $Y^{3+}@C_{82}^-$).[10]

One-electron oxidation of metallofullerenes is carried out when organofluorine radical $^{\bullet}C(CF_3)_2C_6H_4F$ was used as an oxidizing agent.[11] Heating of La@C_{82} and $[FC_6H_4C(CF_3)_2]_2$ solution in 1,2,4-trichlorobenzene to 310 K leads to disappearance of EPR spectra of La@C_{82} over 10 min. Because radical $^{\bullet}C(CF_3)_2C_6H_4F$ itself does not undergo addition to fullerenes, a reaction of oxidative fluorination of metallofullerenes takes place in this case as shown in scheme:

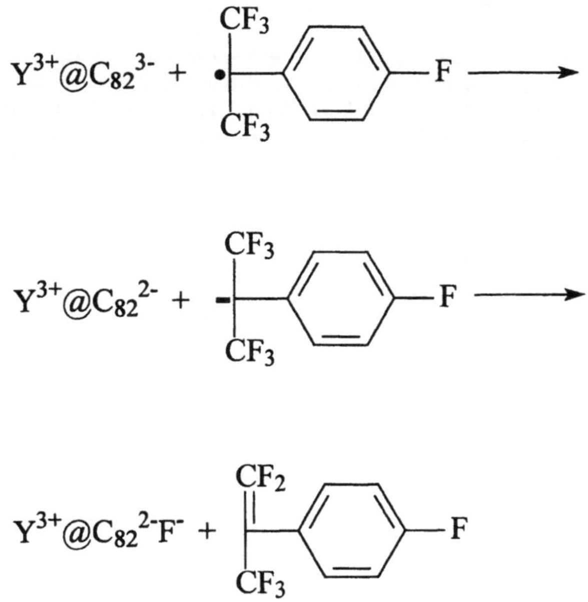

One-electron electrochemical reduction of $M@C_{82}$ leads to diamagnetic salts $[M@C_{82}]^-$ that were isolated and characterized.[12] A dissolution of metallofullerenes in donor nitrogen-containing solvents is accompanied by their reduction.[13]

References

1. D. S. Bethune, R. D. Johnson, J. R. Salem, M. S. de Vries, C. S. Yannoni, *Nature*, 1993, **366**, 123.
2. J. H. Weaver, Y. Chai, G. H. Kroll, C. Jin, T. R. Ohno, R. E. Haufler, T. Guo, J. M. Alford, J. Concaicao, L. P. F. Chibante, A. Jain, G. Palmer, R. E. Smaley, *Chem. Phys. Lett.*, 1992, **190**, 460.
3. R. D. Johnson, M. S. de Vries, J. R. Salem, D. S. Bethune, C. S. Yannoni, *Nature*, 1992, **355**, 123.
4. S. Bandow, H. Shinohara, Y. Saito, M. Ohkohchi, Y. Ando, *J. Phys. Chem.*, 1993, **97**, 6101.
5. K. Kobayashi, S. Nagase, *Chem. Phys. Lett.*, 1998, **282**, 325.
6. B. L. Tumanskii, V. V. Bashilov, S. P. Solodovnikov, V. I. Sokolov, *Full. Sci. Technol.*, 1998, **6**, 445.
7. T. Akasaka, T. Kato, K. Kobayashi, S. Nagase, K. Yamamoto, H. Funasaka, T. Takahashi, *Nature*, 1995, **374**, 600.

8. T. Suzuki, Y. Maruyama, T. Kato, *J. Am., Chem. Soc.*, 1995, **117**, 9606.
9. O. G. Kalina, *M.S. Degree Dissertation*, A. N. Nesmeyanov Institute of Organoelement Compounds, Russian Academy of Sciences, Moscow, Russia, 1999.
10. S. P. Solodovnikov, B. L. Tumanskii, V. V. Bashilov, V. I. Sokolov, S. F. Lebedkin, *Russ. Chem. Bull.*, 2001, in press.
11. B. L. Tumanskii, O. G. Kalina, E. B. Yagubskii, V. P. Bubnov, unpublished results.
12. T. Akasaka, T. Wakahara, S. Nagase, K. Kobayashi, M. Waelchli, K. Yamamoto, M. Kondo, S. Shirakura, S. Okubo, Y. Maeda, T. Kato, M. Kato, Y. Nakadaira, R. Nagahata, X. Gao, E. Van Caemelbecke, K. M. Kadish, *J. Am. Chem. Soc.*, 2000, **122**, 9316.
13. S. P. Solodovnikov, B. L. Tumanskii, V. V. Bashilov, S. F. Lebedkin, V. I. Sokolov, *Russ. Chem. Bull.*, 2001, in press.

Chapter 11

Conclusion

Radical chemistry of fullerenes and their derivatives has accumulated virtually all recent developments in contemporary chemistry of homolytic reactions. Let us summarize briefly available by now data on the structure and reactivity of fullerenyl radicals and ion radicals and the processes involving these species.

It was shown by EPR spectroscopy that the unpaired electron in the $^{\bullet}C_{60}R$ radicals is mostly delocalized over five carbon atoms, incorporated in the two six-membered rings adjacent to the C-CR bond. The dimerization of the fullerenyl radicals occurs according to the "head-to-head" type.

The rate constants of addition of alkyl radical to C_{60} are two orders of magnitude higher than those for the addition of radicals to a wide class of monosubstituted unsaturated compounds.

In the EPR spectra of radical adducts of C_{60}, dynamic effects are manifested, which are associated with the exchange of substituents between positions nonequivalent with respect to the unpaired electron orbital. An explanation based on quantum-chemical calculations of charge distribution on the C_{60} surface that indicate that more electronegative ligands are attracted to the regions of more positive charge over pentagon. The addition of several free radicals affords stable fullerenyl radicals of allyl or cyclopentadienyl types.

The number of radical adducts of C_{70} varies from 3 to 5 depending on the electrophilic properties of reacting radicals, one of the adducts can be stable owing to the substantial delocalization of the unpaired electron.

Addition of phosphoryl radicals to [76]fullerene gives six stable regioisomeric radicals of the nineteen possible isomers. Quantum-chemical calculations were used to determine the structure of radical adducts derived from higher fullerenes.

EPR spectra of adducts of phosphoryl radicals with fullerene derivatives have provided information on the reactivity of nonequivalent carbon atoms of distorted sphere of the polyhedron.

Typical EPR spectra of anion and cation radicals of C_{60} and its certain derivatives as well as examples of preparatively significant ion radical reactions have been described. Heating or irradiation of dimers $(C_{59}N)_2$ or $(C_{69}N)_2$ leads to the formation of heterofullerenyl radicals capable of participating in radical reactions.

The paramagnetism of endometallofullerenes $M@C_{82}$ (M = La, Y) is retained upon addition of even number of radicals or upon oxidation to $M@C_{82}^{+2\bullet}$.

Radical reactions were successfully used for the synthesis on a preparative scale of certain fullerene derivatives including dimers of fullerenyl radicals and halogenofullerenes.

Fullerene derivatives have been obtained, whose unpaired electron is localized both on the fullerene framework and on the attached group of atoms. Nitroxyl fullerene derivatives are reduced to give biradicals in which the fullerene-localized electron can interact with the unpaired electron of the nitroxyl part of the molecule.

Reduction of the benzoquinone-linked C_{60} produces the dianion-radical containing the semiquinone-radical and C_{60}-anion moieties. Paramagnetic metal chelates of an *o*-quinone derivative of fullerene with Mn and Re carbonyls were also obtained.

Modification of fullerenes by radical addition of macromolecules can provide polymeric materials with interesting properties and technological applications.

The ability of fullerene to attach a large number of radicals can be widely used in biology and medicine. Water-soluble fullerene derivatives has been examined for their antioxidant effect on prevention of lipid peroxidation induced by superoxide and hydroxyl radicals. The protection effect on lipid peroxidation was found to be high.

In spite of the fact that the radical chemistry of fullerenes is well studied to date, we do not consider this field of fullerene chemistry to be exhausted.

Water-soluble fullerene derivatives form colloid solutions in water, and the reactivity of fullerenes in colloid state toward radicals is unknown yet.

The development of methods for preparing heterofullerenes containing different number of heteroatoms—such as N, B, P, Si etc.—and study of their structure and reactivity will set a great deal on research techniques capable of operating with trace amounts of substance.

In this respect, it is promising to apply the "phosphoryl EPR spectroscopy of fullerenes" consisting in studying the adducts of fullerenes with phosphoryl radicals by EPR spectroscopy that provides an information on the reactivity of heterofullerenes in radical reactions, polyhedron distortion degree, and the number of nonequivalent atoms, as well as spin density distribution within the radical adducts.

Further studies in the domain of ferromagnetism of fullerene complexes and their polyradical derivatives will be continued. Fullerenes will be widely used as spin traps for studying mechanism of radical reactions both in solution and in gas phase. Reactions involving fullerene cation radical as intermediate will be further studied.

The radical reactions of nanotubes is still unexplored. The presence of fullerene-like apices in nanotubes allows addition of free radicals including macromolecular ones. Such a functionalization of nanotubes can modify their properties and open new frontiers of application.

The addition to fullerene of several B-carboranyl radicals is of special interest because it provides the possibility of preparing unique targets for neutron therapy containing tens of boron atoms.

To implement this project, a carborane moiety should be linked to water-soluble functional groups. This is supposed to provide water solubility also for polyadducts of such carboranyl radicals to fullerene.

Fullerene is known as a radical sponge. If the addition of free radicals to fullerene is reversible, the polyadducts of radicals to fullerene would be initiators of radical processes or mediators of controllable radical polymerization. However, alkyl radicals undergo irreversible addition to fullerene. The strength of C_{60}–heteroradical bond in O-, S-, P-, Si-, Sn-centerd radicals is less than C_{60}-C strength and their heating may result in decomposition of the radical adducts. The initiation of monomer polymerization on account of thermal decay of heteroatomic fullerene derivatives can produce new fullerene-containing polymers.

Thus, we have demonstrated important fundamental and applied significance of radical reactions for fullerenes. We hope this monograph to attract attention of both researchers interested in the problems of up-to-date chemistry of homolytic reactions and those who engaged in studying fullerenes.

Index

Developments in Fullerene Science

Series Editor:

Tibor Braun, *Institute of Inorganic and Analytical Chemistry, L. Eötvös University, Budapest, Hungary*

1. T. Braun (ed.): *Nuclear and Radiation Chemical Approaches to Fullerene Science.* 2000 ISBN 0-7923-6524-0
2. B. Tumanskii and O. Kalina: *Radical Reactions of Fullerenes and their Derivatives.* 2001 ISBN 1-4020-0176-2

KLUWER ACADEMIC PUBLISHERS – DORDRECHT / BOSTON / LONDON

MIX
Papier aus verantwortungsvollen Quellen
Paper from responsible sources
FSC® C105338

If you have any concerns about our products,
you can contact us on
ProductSafety@springernature.com

In case Publisher is established outside the EU,
the EU authorized representative is:
Springer Nature Customer Service Center GmbH
Europaplatz 3, 69115 Heidelberg, Germany

Printed by Libri Plureos GmbH
in Hamburg, Germany